Assessing sanitary mixtures in East African cities

Assessing sanitary mixtures in East African cities

Sammy C. Letema

Environmental Policy Series – Volume 6

Wageningen Academic
P u b l i s h e r s

The research for this PhD thesis was performed
at the sub-department of Environmental
Technology (Prof Jules B. van Lier, promotor)
in cooperation with the Environmental Policy
Group (dr Bas van Vliet, co-promotor) at
Wageningen University.

ISBN: 978-90-8686-215-3
e-ISBN: 978-90-8686-769-1
DOI: 10.3920/978-90-8686-769-1

First published, 2012

© Wageningen Academic Publishers
The Netherlands, 2012

Preface

Many intervention measures have been sought in the past to address sanitation challenges, but largely the approaches have been characterised by a clash between centralised and a decentralised approach. Sustainability assessment has also been characterized by a dualistic approach, between the techno-centric and the eco-centric. The dualistic dichotomy in sanitary provision is often reduced to competition between the proponents, none of them providing a panacea for solving complex sanitary challenges. Such opposing views are often simplistic and not in tandem with the existing multiple sanitary options and service providers in East African cities, which defy such classification. This thesis, therefore, provides the theoretical and empirical basis for a third way of classifying and assessing the multiple technical and institutional options to sanitary provision. This novel assessment approach called 'modernised mixtures' is used in this thesis as a tool for conceptualising, assessing and improving sanitary provision in East African cities. The approach is an inclusive way of assessing sanitary mixtures, which benefits decision making among imperfect options.

The research described in this thesis was financially supported by Partnership Research on Viable Urban Environmental Infrastructure Development in East Africa (PROVIDE), a project funded by Interdisciplinary Research and Education Fund (INREF), Wageningen University, The Netherlands.

I thank the PROVIDE team, Spaargaren, Oosterveer and van Buuren, who came to East Africa in 2006, for showing confidence and picking me as a PhD candidate under PROVIDE project.

I am grateful to Jules van Lier, my promoter, for his support and flexibility in the direction of research and write up of the thesis. My sincere gratitude goes to my co-promoter, Bas van Vliet, for his guidance, belief and inspiration. My appreciation goes to Joost van Buuren for his technical advice and sharing experiences.

My appreciation goes to administrators Corry Rothuizen of Environmental Policy and Liesbeth Kesaulya of Environmental Technology for their administrative and logistical support during my entire PhD period.

My appreciation to my wife Elizabeth and children for their moral support and deep understanding when I was away most of the time.

My appreciation also goes to my PROVIDE colleagues: Aisa, Christine, Fredrick, Judith, Mesharch, Tobias, and Richard, and PROVIDE affiliate, Leah. Being with you people made my stay in Wageningen home away from home.

During the course of my stay in Wageningen I made great friends: Gabor and Lina (Environmental Technology), Marjanneke, Dorien, Carolina, Jennifer, Elizabeth, Natapol, Kim, Harry and Hilde (Environmental Policy), Maren (Development Economics), Bing (Logistics), Nelly (Socio-Spatial Analysis) and Kibor (Plant Sciences). The get-together, dinners and lunches are highly appreciated. Marjanneke, I will always remember the Fellowship of the Movies and Nelly your company to church and bicycle rides.

Table of contents

Chapter 3.
Assessment of urban sewers and treatment facilities in Kampala and Kisumu as interplay of flows, networks and spaces

Chapter 4.
Potentials of satellite sanitary systems in Kampala City

Abbreviations

AWSB	Athi River Water Service Board
CBOs	Community Based Organisations
DAWASA	Dar es Salaam Water and Sanitation Authority
IDAMCs	Internally Delegated Area Management Contracts
IETC	International Environmental Technology Centre
JICA	Japanese International Corporation Agency
KCCA	Kampala Capital City Authority
KIWASCO	Kisumu Water and Sewerage Company
KWP	Kenya Water Partnership
LAs	Local Authorities
LVEMP	Lake Victoria Environmental Programme
LVSWSB	Lake Victoria South Water Service Board
MAPET	Manual Pit Emptying Technology
MCA	Multi-Criteria Analysis
MCK	Municipal Council of Kisumu
MDGs	Millennium Development Goals
MM	Modernised Mixtures
MMA	Ministry of Municipal Affairs
MWE	Ministry of Water and Environment
MWI	Ministry of Water and Irrigation
NEMA	National Environment Management Authority
NGOs	Non-Governmental Organisations
NHCC	National Housing and Construction Company
NWSC	National Water and Sewerage Corporation
PROVIDE	Partnership Research on Viable Environmental Infrastructures in East Africa
PRC	Performance Review Committee
SAs	Sewerage Agencies
SPs	Service Providers
SPA	Service Provision Agreement
STPs	Sewage Treatment Plants
UNICEF	United Nation International Children Education Fund
WASREB	Water and Sanitation Regulatory Board
WCED	World Commission on Environment and Development
WHO	World Health Organisation
WSBs	Water Service Boards
ZPCs	Zonal Performance Contracts

Chapter 1.
Introduction

1.1 Background

The rapid rate of urbanisation in developing countries has created an overwhelming demand for housing, infrastructure and services (Taylor & Parkinson, 2005), whereas sanitary provision is lagging behind urbanisation rates. Many intervention measures have been sought in the past to address water and sanitation challenges, but globally, 2.6 billion people still lack access to improved sanitation (UNDP, 2006). United Nation's member countries in March 1977 declared the period 1981-1990 the International Drinking Water Supply and Sanitation Decade. Despite the concerted efforts during that period, the number of people not served by adequate and safe water supply fell by approximately 450 million whereas those without appropriate means of excreta disposal remained almost the same (Loetscher, 1999; WHO, 1992). Member countries of the United Nations once again met at the turn of the millennium and agreed on Millennium Development Goals (MDGs), where they set, among others, a target of halving by 2015 the proportion of people without access to improved sanitation and significantly improving the lives of slum dwellers by 2020. Besides, WHO and UNICEF have set a target of 'Sanitation for All' by 2025. To achieve WHO/UNICEF target, 480,000 people would have to be provided with improved sanitation daily (Mara, Drangert, Anh, Tonderski, Gulyas, & Tonderski, 2007). To give more impetus to the magnitude of the sanitation challenge, 2008 was declared the 'International Year for Sanitation' by United Nation's General Assembly in December 2006. There is, however, dismal progress towards achieving the targets in most Sub-Saharan Africa countries, East Africa included, thus are unlikely to meet the MDG target (UN, 2006; UN-Habitat, 2008; WHO/UNICEF, 2010). Despite years of intervention measures between the Water and Sanitation Decade and the International Year of Sanitation, the proportion of population using improved sanitation in Sub-Saharan Africa increased only marginally from 28% in 1990 to 31% in 2008 (WHO/UNICEF, 2010).

To achieve adequate and sustainable urban sanitary services, a proper institutional framework, adoption of appropriate technology and embedding of sanitation solutions in local socio-economic, cultural and spatial structure is imperative (WECD, 1987; Ellege, Myles, & Warner, 2002; Seghezzo, 2004). Technologies are considered appropriate when they fit in the boundary conditions determined by local conditions. Such boundary conditions consist firstly of standards and principles of engineering, which determine the way in which sanitary systems develop. Interestingly, most boundary conditions for sanitary services follow a conventional master plan of city development geared towards centralised systems and availing of planned and serviced land for new settlements to all city residents. Secondly, most regulations, institutions and organisational frameworks for sanitary provisions are public oriented and in line with the engineering master plan. Yet there are multiple providers of sanitary services in the rapidly developing cities. Thirdly, different spatial structures have different affinities for particular sanitary systems. So far centralised sanitary systems, comprehensive urban planning, and public provision in developing countries, especially East Africa have had little impact, as about 50-70% of urban population live in informal settlements that are neither planned nor serviced (UN-Habitat, 2003, 2008). Fourthly, socio-economic and cultural conditions, i.e. affordability, acceptability and accessibility; determine the feasibility of sanitation options for adoption.

The picture of development efforts towards improved sanitary provision in East African cities is plagued by contradictory development strategies pursued by many agencies with relative degree of autonomy. Consequently, there is lack of a cohesive and wholly accepted strategy for sanitary provision in cities of East Africa due to the co-existence of various sanitary solutions, spatial structures and multiple providers resulting in sanitary mixtures. To reach the targets as formulated in the MDG and WHO/UNESCO, sanitary mixtures seem to offer better impetus and hence there is a need for a new approach and tools to assess and evaluate existing sanitary mixtures. We utilise a sanitary mixtures approach in this thesis to move away from dichotomy of centralised systems that are often referred to as conventional or modern on the one hand and decentralised systems that are often termed traditional or onsite on the other. Instead we aim at adoption of locally embedded solutions that merge the best option of both conventional and traditional systems in fitting local conditions and that comply with sustainability criteria of public and environmental health, accessibility and flexibility.

This thesis posits that assessment of sanitary mixtures, as is the case in contemporary East African cities, can best be achieved by assessing sanitary systems on two levels. Firstly, by

Table 1.1. Poverty, slum and urban sanitation coverage in East Africa countries.

%	Population below income poverty line (UNDP, 2009)		Population in slum (UN-Habitat, 2008)	Sanitation coverage in 2008 (WHO/UNICEF, 2010)			
	Maximum income 2$/day 2000-2007	National poverty 2000-2006	2005	Improved[1]	Shared[2]	Unimproved[3]	Open defecation[4]
Kenya	40	52	55	27	51	20	2
Tanzania	97	36	66	32	30	36	2
Uganda	76	38	67	38	56	4	2
Rwanda	90	60	72	50	18	31	1
Burundi	93	68	65	49	22	27	2

[1] Improved sanitation: flush/pour-flush toilets connected to piped sewer systems, septic tanks or pit latrines; ventilated improved pit (VIP) latrines; pit latrines with slabs; composting toilets.

[2] Shared sanitation: using a public facility or sharing any improved facility.

[3] Unimproved sanitation: the collective name for sanitation solutions as flush/pour flush toilets without proper connections; pit latrines without slabs or open pits; buckets and hanging toilets/ latrines.

[4] Open defecation: no facilities are present and the surroundings (bush or field) are used for excretion.

assessing sanitary options along four provision dimensions: scale, management, flows and end-user participation. Secondly, by assessing the sustainability of the sanitary options based on the three criteria: public and environmental health, accessibility and flexibility. This assessment tool making use of the mentioned dimensions and the sustainability criteria is further referred to as the modernised mixtures (MM) approach (Oosterveer & Spaargaren, 2010; Van Vliet, 2006). Following this approach, sanitary mixtures are considered modernised when they better fit the local physical and socio-economic systems regarding scales, strategies, technologies and decision making structures.

1.2 Urbanisation and sanitation provision status in East Africa

Although East Africa is the least urbanised African region, it is experiencing rapid urbanisation exceeding 3.9% annual growth between 2000 and 2015 largely due to natural growth devoid of basic infrastructures (UN-Habitat, 2008). The urbanisation, however, is not accompanied by industrialisation, economic growth, spatial planning or investment in environmental infrastructures, leading to urbanisation of poverty and the growth of extensive informal settlements (Table 1.1). The urbanisation of poverty poses a threat to environmental health, perpetuates social exclusion and inequalities, and creates service gaps (UN-Habitat, 2008).

Different sanitary approaches attributed to parallel sanitary solutions pursued under different intervention measures culminate into various stages of sanitary solutions, which all sit next to each other. The mixture comprises of different sanitary systems having different coverage, quality and scale (Tables 1.2 and 1.3), different institutional arrangements, and servicing of different urban spaces and clientele. The number of urban centres connected to modern sewerage accounts for about 14% in Kenya, 12% in Uganda excluding Town Boards, 16% in Tanzania and none in Rwanda and Burundi. Those that have modern sewerage, however, have a low coverage, ranging from 5-36% in Kenya, 0.9-20% in Tanzania and 2-26% in Uganda (Tables 1.2 and 1.3). The coverage and connection ratio are not generally in tandem with water supply coverage.

The status of sewage treatment works is disappointing. For instance, in Kenya, out of 38 sewage treatment plants (STPs), 40% are overloaded, 15.5% are operating at design capacity, 2.5% are not

Table 1.2. Population coverage (%) of sanitation solutions in East African capital cities.

City	Sewerage	Septic tank	VIP latrine	TP latrine	No facility	Reference
Nairobi	36	<·················· 64 ················>			n.a	(AWSB, 2005)
Kampala	6	18	<······ 70 ········>		6*	(NWSC, 2004)
Dar es Salaam	13	13	n.a.	70	4	(DAWASA, 2008)
Kigali	0	16	3	80	1	(Sano, 2007)
Bujumbura	0	n.a.	n.a.	99	1	(WSSINFO, 2008)

Abbreviation: n.a. not available; * 3% practice open defecation and 3% other sanitation options; VIP ventilated improved pit; TP traditional pit

Table 1.3. Sewer coverage, areas and treatment systems in East Africa.

Tanzania		Uganda		Kenya	
Urban centre/ sewer areas	% P coverage	Urban /sewer area	% P coverage	Urban/sewer areas	STPs
Arusha	9	Entebbe$_2$	4	*Naiorbi$_{20}$, Kisumu$_3$, Mombasa$_3$*	Conventional
Dar es Salaam$_{10}$	13.6*	Fort Portal	2	*Kitale$_2$, Eldoret$_2$,* Kericho	trickling filters
Dodoma	20	Gulu	7	Kiambuu, Limuru, Webuye,	Oxidation ditch
Iringa	5	Jinja	22	Naivasha	
Kilimanjaro	9	Kabale	11	Athi River, Eldoret, Embu,	Waste
Kigoma	-	Kampala$_{10}$	5	Homa Bay, Meru, Machakos,	stabilisation
Mbeya	-	Lira	2	Nanyuki, Ngong, Nyahururu,	ponds
Morogoro	0.87	Masaka	8	Nairobi, Voi, Nakuru, Isiolo,	
Mwanza	9	Mbale	26	Kisii, Nyeri, Thika, Busia,	
Tabora	3.9	Mbarara	5	Kapsabet, Karatina, Kakamega,	
Tanga	15.5	Soroti	2	Kericho, Bungoma, Kisumu,	
		Tororo	7	Kitale, Muranga	

Symbol: *Italics* towns with more than one sewerage area; subscript show the number of sewerage areas; P population; STPs sewage treatment plants. Sources: Chaggu, 2004; *DAWASA, 2008; NWSC, 2009; MWI, 2008a.

operating at all and 42% are operating below capacity (MWI, 2008a). The treatment systems for conventionally collected sewage using gravity sewers are mostly waste stabilisation ponds (WSPs), a few of them with mechanised processes such as conventional trickling filters, oxidation ditches and aerated lagoons. In Kenya, WSPs are used in 25 out of 38 urban STPs. In Uganda, 12 are WSPs whereas 2 are conventional trickling filters based STPs. Stringent environmental standards set by National Environment Management Authority (NEMA) needs, in addition to carbon and pathogens, also nutrient removal, which makes most conventional treatment process options not in compliance with effluent discharge standards. The very stringent legislation on the one hand and the socio-economic inabilities to meet the set requirement using up to date technologies on the other hand, paralyses any investment at the wastewater treatment level. Here a paradigm shift is urgently needed.

1.3 Variety of sanitary scales and institutional arrangements

Sanitary provision in East African cities are a patchwork of systems: conventional urban sewers connected to STPs, satellite sewers with decentralised treatment and the onsite systems, e.g. septic tanks, ventilated improved pit (VIP) latrines and traditional pit (TP) latrines (Tables 1.1-1.3). In Nairobi, there are 20 sewerage systems servicing about 36% of the population: 5 urban public sewerage areas owned by Athi Water Services Board (AWSB) and operated by Nairobi Water and Sewerage Company. The remaining 15 sewerage systems are privately owned and operated satellite systems. There are 4 servicing security forces (army, police and prison), 5 colleges and universities, 3 schools, 2 industries, and 1 servicing a hotel. In Kampala, there is one urban public sewerage system owned by National Water and Sewerage Corporation (NWSC) and operated by Kampala Water Partnership (KWP) and 9 privately owned satellite systems. Four of them are servicing residential areas, 3 college/universities, and 2 are servicing security forces (police and prison). In Dar es Salaam, there is one urban public sewerage system owned and managed by Dar es Salaam Water and Sanitation Company (DAWASA) and 9 satellite systems. Three of them are servicing defence and security forces, 2 industrial, 1 a university, 2 residential areas and 1 servicing an airport. Besides, there are also septic tank, VIP latrine and TP latrine systems (Table 1.2) that are mainly provided by households, communities, voluntary sector and quasi-public institutions. Urban sewerage systems are often centralised and large-scale in approach whereas onsite systems are decentralised, at household and community scale and perceived as small-scale. Satellite systems are semi-collective systems and thus can be perceived as neither centralised nor decentralised. Sanitary systems, therefore, are different in terms of technical scales and institutional arrangements, which may have implications on the level of end-user participation, nature of sanitary flows and the way management interventions, can be made. Such differences may offer impetus for selection of locally embedded sanitary options in towns and cities.

1.4 Research objectives and questions

Achieving the MDG goal of halving by the year 2015 the number of people without adequate sanitation or WHO/UNESCO sanitary for all by 2025 in East African cities is a daunting task. The diversity of implemented sanitation solutions (Tables 1.2 and 1.3) has led to sanitary mixtures. Conceptualising and assessing sanitary mixtures requires new conceptual models instead of the espoused centralised or decentralised approaches. This thesis, therefore, utilises the earlier mentioned MM approach in assessing sanitary mixtures in East African cities taking Kampala (Uganda) and Kisumu (Kenya) as case study cities. To achieve this, the four objectives formulated for this thesis are to:

1. Make an inventory of sanitary systems in Kampala and Kisumu.
2. Assess and map sanitary systems configurations along MM dimensions in Kampala and Kisumu.
3. Assess sustainability of sanitary systems on MM criteria in Kampala and Kisumu.
4. Enhance insights on the usefulness of using the MM approach as conceptual model and as an assessment and prescriptive tool for sanitary mixtures in East African cities.

In order to achieve the research objectives, the following research questions are formulated for the thesis:

1. What are the types of sanitary systems in Kampala and Kisumu?
2. What are the configurations of sanitary system in terms of MM dimensions in Kampala and Kisumu?

3. To what extent are the sanitary systems in Kampala and Kisumu considered sustainable following MM sustainability criteria?
4. To what extent does the MM approach provide a useful conceptual model and a tool for assessing, prescribing and generalising on sanitary systems in East African cities?

1.5 Operationalization, limitations and methodology

The strategy chosen for this thesis research is a case study approach. A case study is a strategy for description and explanation of group attributes, patterns, structures, and processes over time and space through strategic selection and comparison of a few cases and sub-cases (Gray, 2004; Verschuren, 2002; Yin, 1984; Zhang, 2002). One way of choosing cases is by utilising typologies (Silverman, 2000). From the typology of primary and secondary cities in Table 1.4, primary cities in Kenya, Uganda and Tanzania as well as secondary cities in Kenya have urban sewer systems since colonial time. However, none of the secondary cities in Tanzania and Uganda had urban sewer system during colonial time. Strikingly, neither the primary nor the secondary cities of Rwanda and Burundi has an urban sewer system. From the typology in Table 1.4, a primary city in Uganda (Kampala) and a secondary city in Kenya (Kisumu) are chosen for case studies. The two cities are assumed to offer rich cases and to be representative of other East African cities. Both Kampala and Kisumu cities discharge their wastewater into Lake Victoria, with Kisumu being the headquarters of East Africa Lake Basin Commission and Kampala being a primary city in Lake Victoria region. Both cities are in synchrony in terms of sewerage development in the 1930s and at the turn of 21st century and over the next two decades as espoused in their sanitary master plans.

Sanitary scale, settlement structure and institutional arrangements are utilised to stratify, purposively sample and study sanitary systems at three levels: urban, satellite and onsite. The MM approach is used as a conceptual model and assessment tool. Data collection techniques entailed document acquisition, archival retrieval, interview schedules, observations guides and wastewater sampling. Data analysis is done through content analysis for qualitative data and descriptive statistics for quantitative data. Triangulation is applied as a method for quality control and validation of data by blending quantitative with qualitative data and field survey with desktop studies. This is achieved by triangulating interviews and observation with documentation and archival retrieval and practices with standards and guidelines. The study assesses sanitary dimensions on four axes, i.e. technical scale, management scale, level of flow separation and level of end-user participation and on 6 levels, and assesses sustainability on three criteria, i.e. public and environmental health, accessibility and flexibility based on 34 indicators. Assessment indicators

Table 1.4. Typology of cities with sewerage system since colonial time in East Africa.

City characteristics	Kenya	Uganda	Tanzania	Rwanda	Burundi
Primary city	1	1	1	0	0
Secondary city	3	0	0	0	0

for each criterion are developed through literature review. A multi-criteria analysis (MCA) is used in the assessment of sanitary systems by way of means scores and weighted mean scores and results presented in a matrix and figures. Although Kampala and Kisumu cities are unique in many respects, generalisation can be made towards East African cities as a whole in terms of the variety of sanitary solutions. Besides, the configurations from the cases can, to a certain degree, be discernible and thus applicable across East African cities.

1.6 Thesis structure

This thesis presents results of primary data from fieldwork and analysis of secondary data in Kampala (Uganda) and Kisumu (Kenya) on sanitary provision dimensions and sustainability assessments. The thesis is structured into seven chapters. Chapter 2 handles the theoretical basis within which a shift from centralised and decentralised approach to sanitary mixtures that are considered modernised is underpinned. Theories of modernisation, sanitary provision dimensions and assessment scales are discussed. Chapter 3 presents the status of conventional urban sewers and treatment facilities in Kampala and Kisumu in terms of flows, networks and spaces against spatial structure and institutional arrangements. This chapter not only informs about the paradigm within which urban sanitary policy was structured in the last century, but also assesses their provision dimensions. Chapter 4 presents the characteristics, status and potential of satellite sanitary systems as an intermediate sanitary provision pathway in East African cities, taking Kampala as a case. Chapter 5 looks at onsite sanitary provision and their challenges using a chain approach. It also examines onsite systems as permanent or transient solution depending on density, spatial requirements and excreta flow, besides assessing their configurations. Chapter 6 assesses the sustainability performance of six sanitary systems studied in Chapters 3-5 on three MM criteria. Finally, Chapter 7 presents the summary of findings, generalises and reflects on the applicability of MM approach as an assessment and prescriptive tool, and makes a conclusion.

Chapter 2.
Modernised sanitary mixtures: a paradigm shift in sanitary provision

2.1 Introduction

Different sanitary infrastructures exist in cities of developing countries like East Africa attributed to parallel development of sanitary solutions pursued by different actors under different intervention rationalities. Different intervention measures culminate in sanitary solutions of varying modernities, which all sit next to each other in the city. The result is a sanitary mixture and the challenge is which approach to adopt and improve the mixture. Such mixtures require a structured approach to study, assess and configure them to meet societal needs and enhance their sustainability. This chapter, therefore, positions the development direction of current sanitary mixtures within a modernisation discourse as a strategy for providing sustainable sanitary mixtures in East African cities.

2.2 Modernisation and modern infrastructural ideal

Modernisation, generally, is a loosely used term, which means different things to different people. On the one hand, it is a historical period that occurs in phases leading to modernities which, in Western societies, are pre-modern, modern or post/late modern (Arts, Leroy, & Van Tatenhove, 2006; Castells, 1996). Each modernity period is rationalised and universalised, with the assumption that programmes of modernity and their institutional arrangements would prevail in modernising and modern societies (Eisenstadt, 2000). Spaargaren (2003) conceptualises models of infrastructure, and implicitly sanitary provision within Western modernisation periods as stand-alone, centralised and fragmented networks reminiscent of early modern, modern and late-modern period, respectively. However, Van Vliet (2005) notes that even within Western societies, infrastructures have not evolved in a linear pattern but as different infrastructural modes of provision classified as autonomous, piecemeal, integrated, universal and marketed, which can coexist at different contexts in varying degrees. The sanitary modernisation era, 1870-1970, introduced and institutionalised modern management to Western cities: public provision of infrastructure and services through formal departments and utility agencies, application of scientific expertise to resolve sanitary problems, professionalization of services through civil service technocrats such as planners, civil engineers, sanitary specialists, and a centralised approach to water supply and wastewater management (Pincetl, 2010). Graham and Marvin (2001) also note that cities were reorganised to meet a sanitary city ideal through standardised roads, water supply, wastewater collection and treatment, energy provision and communication, which embody universal coverage. Standardisation, in essence, entrenched public provision of environmental infrastructures and services as the norm, a conventional way of city infrastructure provision by technocratic public agencies centred at municipalities providing a bundle of municipal services as public or merit goods. Centralisation of infrastructures and services at municipal level was an attempt to coordinate, realise economies of scale, standardise technologies and services, and attain efficiency. Existence of multiple modes of infrastructure and services provision challenges the infrastructural modernism structures that were spread across the world through transfer of technology and institutional frameworks applied in Western societies.

Western modernisation and resultant modernities and their structures of service provision have not resonated well in developing countries. Consequently, other alternative theories have emerged challenging Western ideal of modernity and offer alternative modernisation pathways. One such alternative to Western derived modernisation is multiple modernities (Eisenstadt, 2000), which embody acceptance of multiple rationalities, diversity and multiplicity that disputes a universal approach to modernity (Harrisson, 2006; Lee, 2006, 2008).

Colonisation introduced and entrenched Western modernism, especially technological standards and choices, engineering codes and principles, and institutional arrangements. Modernisation in developing countries cities, however, should be seen on the one hand as imposition of Western copies of technologies, economic and institutional models during colonial period (Eisenstadt, 2000), and on the other hand, as selective incorporation of technologies, discourses, and institutions of the Western modernity during post-colonial period to create a distinct form of modernity argued by Hancard (2001) in Harrisson (2006). In essence, therefore, developing countries are not just copying Western modes of modernisation but undergo a process of selection and appropriation. This process of selection and appropriation, however, results in various forms of altered modernities. Altered modernities move us away from adopting the common notion of failed, incomplete or deteriorated modernity in developing countries (Harrisson, 2006), to viewing them as a breeding ground for not only alternative but also multiple modernities. Multiple modernities take us to viewing the development of a variety of systems not only as different entities, but also can form integrated systems in the city for sustainable development. This is particularly the case with sanitary provision in developing countries, and East African cities in particular, which experience a diversity of sanitary solutions and multiple providers. There is need, therefore, for assessment tool for such diverse and mixed sanitary solutions.

2.3 Spatial-technical dimensions of sanitary provision

2.3.1 Paradigms of centralisation and decentralisation

Technology development framing in the context of service provision of water, waste (water), energy, housing and food over the last five decades has been characterised by a clash between two paradigms: a centralised or conventional approach and a decentralised or alternative approach. The latter can also be regarded traditional as they were applied before the centralised or conventional approach was regarded as modernity. Centralised systems are viewed by the proponents of alternative approach as large-scale, centralised, expert driven, complex, and ecologically unsound whereas the alternatives are small-scale, decentralised, participatory, simple and ecologically sound (Figure 2.1) (Smith, 2005). The proponents of centralised systems argue that they have provided hygienic conditions, easy transport with little visibility, adequate handling of organic matter and nutrients, and with little energy consumption (Harremoës, 1997). Moreover, low-tech craft

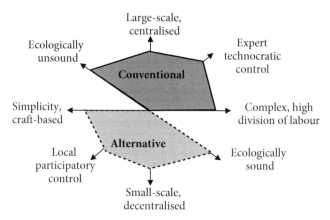

Figure 2.1. Sanitary systems classification in 'conventional' and 'alternative' along multidimensional axes (adopted from Smith, 2005).

systems are not necessarily sustainable at any costs and design, whereas appropriateness depends on local conditions (Grau, 1996).

On the other hand, it is argued that conventional systems have inertia and lock-in effects, which curtail emergence of alternative decentralised options on house-onsite and community level to develop and complement them (Hegger, 2007; Nilsson, 2006; Van Vliet, 2002, 2006). Strikingly, so far, centralised versus decentralised debates, as argued by Bijker (1995) as cited in Smith, (2005), are often reduced to competition between the proponents in an attempt to remain relevant, retain, access, or wrestle power, but with each group possessing various, but always incomplete levels of capital, scientific expertise and technology.

Urban systems for waste(water) often develop in a paradigmatic manner, where certain engineering practices, standards, and technical knowledge come to prevail, which may deter technological changes (Chartzis, 1999; Ertsen, 2005 in Nilsson & Nyanchaga, 2008). Conventional sewerage is based on conservative design values that have undergone little changes over a century. For smooth operation, the resulting gravity-based systems require high water flows, minimum pipe diameters, high number of household connections, sewerage passing both sides of the street, minimum velocity, minimum depth and slope of sewers, pumping stations at various stages of the sewer network, and design periods of over 30 years (IETC, 2002; Mara, 1996; Mara & Alabaster, 2008; Paterson, Mara, & Curtis, 2007; Sundaravadivel, Doeleman, & Vigneswaran, 1999). The applied conservative design values result in deep sewerage, high capital costs, high operation and maintenance efforts, and inappropriateness in most types of urban settlements (IETC, 2002; Otis, 1996; Sundaravadivel *et al.*, 1999).

Conventionally designed urban sanitary systems comprise of large-scale sewer collection and treatment systems characterised by large piping networks that convey wastewater from the site of generation to the site of treatment, making use of pumping stations and complex siphons. Moreover, such systems are dependent on advanced water supply and electricity infrastructures in place, and towns engineered into pipe like networks (Graham & Marvin, 2001; Newman, 2001; Van Lier & Lettinga, 1999). Large-scale systems, Van Dijk (2008) notes, are too expensive to introduce on a large-scale in developing countries. Consequently, currently existing large scale systems serve only a small population, are capital intensive in development and maintenance, and subsidises the more affluent groups (Nilsson, 2006; Oosterveer & Spaargaren, 2010; Toubkiss, 2010).

Alternative traditional sanitary options are the onsite systems coupled with offsite treatment of manually collected wastes or *in situ* waste valorisation linked with reuse practices. Onsite systems, e.g. pit latrines and septic tanks are cheapest and most appropriate for rural, low-density urban and low-incomes areas; and can provide the same health benefits and user convenience as conventional sewerage systems provided ground water is deep and areas are not prone to flooding (Kalbermatten, Julius, & Gunnerson, 1982; Paterson *et al.*, 2007). Construction and management of traditional onsite systems such as latrines is well described in text books (e.g. Franceys, Pickford, & Reed, 1992). Although developed for rural, low-density applications, onsite sanitary systems serve the majority of the urban population in developing countries, offering solutions to individual or group of households and accounting for over 80-100% of sanitation solutions in cities (Kone, 2010). Onsite sanitary systems are stand alone, site specific, individual plot-based, and very basic options that are often temporary facilities (Abbott, 2010). However, onsite systems are perceived a simple and second best option, useful in situations where the finances, technological capabilities and organizational capacities are severely limited for centralised systems. Besides, they are supposed to be transient, i.e. replaced with more advanced systems as soon as the social, economic and technological conditions allow (Spaargaren, Oosterveer, Van Buuren, & Mol, 2005). Although often implemented, however, they are not feasible in (peri-) urban areas due to high population densities, lack of space, poor drainage and risk of water sources contamination (Paterson *et al.*, 2007).

Besides the conventionally designed centralised and traditionally developed onsite systems, there are other intermediate options, which are twofold. Firstly, there are the simplified sewerages, e.g. condominial and settled sewer systems coupled with offsite treatment. Simplified sewers have emerged with relaxed designs codes. Such sewers result in use of small-sewer pipe, reduction in water requirements, lower gradients and depth, and manholes replaced by inspection chambers or cleanouts; while maintaining sound design principles (IETC, 2002; Mara & Alabaster, 2008; Paterson *et al.*, 2007; Reed, 1995; Sundaravadivel *et al.*, 1999). Alternative sewers are viewed to be low-cost, flexible in location and layout, amenable to, and frequently even dependent on, community participation, appropriate for planned and unplanned settlements, and can be planned as decentralised networks, utilising low-cost treatment (Mara & Alabaster, 2008; Paterson *et al.*, 2007; Pombo, 1996; Sundaravadivel *et al.*, 1999). Secondly, there are the autonomously functioning satellite sewers and treatment systems, which are serving a designated city section, often covering part of a catchment in which solely gravity sewers can be used. Such intermediate semi-collective sewerage and treatment systems, serve clusters, communities and neighbourhoods (Gómez-Ibáñez, 2008; Hunt, 2005; Mara, 2008; Toubkiss, 2010). Various authors claim that intermediate infrastructures have a number of advantages (Gómez-Ibáñez, 2008; Hunt, 2005; Kariuki & Schartz, 2005; Toubkiss, 2010), since they increase access to sanitary services without being dependent on large scale infrastructural works and institutional support. In various cases the private sector is involved in both sewage collection and treatment.

2.3.2 Scale

Sanitation scales are highly contestable. A first distinction can be made between technical and management scales (Van Vliet, 2004, 2006). Other distinctions based on the amount of inflow to STPs in terms of population served and urban spatial planning hierarchy (De Graaf, 2006; Hasselaar, 2006; Hegger, 2007; Mgana, 2003; Rijnsburger, 1996; Van Buuren, 2010) (Table 2.1). However, there is no absolute delimitation of the maximum or the minimum number of users within a scale. Crites and Tchobanoglous(1998) classifies STPs as small scale and decentralised having a treatment capacity of <3,785 m³/d (1 MGD), which is about a population of 30,000. Following Van Buuren's (2010) classification, the maximum capacity of a decentralised municipal system has been set at a population of 50,000 or an area of 250 ha, whereas community sanitary systems process 4,000 m³/d and serve about 20,000 people and occupy a maximum area of 100 ha. This thesis is of the view that scale is locally embedded and different sanitary solutions occupy different technical and spatial scales that depend on the local conditions.

From Table 2.1, six scales can be deduced; dwelling unit, community/group of households, neighbourhood, small-urban, medium-urban and large-urban. Therefore, scale is not only about large or small as often perceived in centralised-decentralised paradigms, but also includes the number of people served. Therefore, there is a range of possibilities between the two.

The type of settlement also determines the scale and type of sanitary option, with different settlements having different affinities for a particular type of sanitary option. Oosterveer and Spaargaren (2010) argue that centralised systems make strong assumptions about homogeneity in housing stock, density, degree of urbanisation, accessibility and related infrastructures. Moreover, it is noted that centralised systems require towns engineered into pipe-like networks while each individual plot be accessible and standardised to attain universal connections and mass consumption (Graham & Marvin, 2001; Newman, 2001; Oosterveer & Spaargaren, 2010; Spaargaren, 2003). East African cities, however, are characterised by large segments of informal

Table 2.1. Sanitation scale and service level based on population and household size.[1]

Reference	Rijnsburger (1996); Mgana (2003)	De Graaf (2006); Hasselaar et al. (2006); Hegger (2007)	Van Buuren (2010)
Sanitation scale & service level	• housing unit 10-40 P pit latrines/septic tanks • housing block 40-200 P or 4-10 Hh and mostly septic tank • neighbourhood unit 100-2,000 P and mostly wastewater collection and treatment	• dwelling 1 Hh • houses/apartments cluster of 2-25 Hh • neighbourhood 25-250 Hh • city quarter 250-10,000 Hh • city or large <10,000 Hh	• individual onsite/cluster 5-50 P • community >50-2,500 P • small-scale >2,500-50,000 P • medium-scale >50,000-500,000 P • large-scale >500,000 P

[1] Hh = household; P = population (no. people).

settlements comprising 30-70% of the population (Kombe, 2005; Olima, 1994; UN-Habitat, 2008; Nawangwe & Nuwagaba, 2002) that are either informal, unplanned or of very low densities. We argue that such diversity of spatial structures calls for provision of different sanitary solutions to suit the local conditions.

2.3.3 Flows

Sanitary flows are negative valued, pathogenic and are an immediate threat to public health. However, and interestingly, sanitary flows are currently not only regarded as negative valued nuisance flows that require extensive treatment but are increasingly considered for recycling and reuse, valorising the wastewater components into useful products (Abu-Ghunmi, 2010; Otterpohl, Albold, & Oldenburg, 1999; Otterpohl, Braun, & Oldenburg, 2003; Zeeman & Lettinga, 1999). Presently, in urban sewerage and drainage, three levels of sanitary flow separation are distinguished: (1) separation of rain or storm water and municipal sewage, (2) separation of industrial wastewater and domestic sewage, and (3) source-separation of domestic wastewater into black, grey and rain water (Abu-Ghunmi, 2010; Kujawa & Zeeman, 2006; Otterpohl *et al.*, 1999, 2003; Van Buuren, 2010). Application of urine diverting toilets in source-separation toilets result in two additional flow streams: yellow water (urine with or without flush water), and brown water (faeces with water) (Kujawa & Zeeman, 2006).

Other aspects of sanitary flows comprise water supply and wastewater generation. The per capita water consumption (l/ca*d) or consumption per hectare (m³/ha*d) determines the kind of sanitary system to adopt. At a base flow density of >10 m³/ha*d, the feasibility of onsite sanitary systems is questionable whereas the construction of sewerage is regarded as more appropriate (Veenstra & Alaerts, 1996 as cited in Chaggu, 2004). In addition, the permissible effluent quality discharge standards to the environment, i.e. linked to organic matter, nutrients and pathogens, will, amongst other factors, determine the nature of treatment system required.

2.3.4 Mixed sanitary solutions

The above analysis reveals that if sanitary provision is to succeed in East African cities, where socio-economic factors limit both capital and operational exploitation costs, it should be based on mixed solutions and at multiple scales. A mixed sanitary structure can be conceptualised spatially as illustrated in Figure 2.2, a parallel development of different systems at different scales serving different parts of the population. Each sanitary system's service level can have its treatment scale and technology option. Adoption of mixed sanitary solutions may introduce complexity, which may lead to increased operation and maintenance costs, personnel and problems due to lack of standardization, non-up to date infrastructure records and weak enforcement. On the other hand, recognition of these mixed solutions give ample possibilities for full coverage of sanitary services provided the offered solutions are meeting agreed sustainability criteria. Therefore, in order to include all available sanitary structures in a strategic urban master plan each applied system and technology requires embracing a defined set of sustainability criteria that will denominate the respective sanitary solution as 'modernised' (Section 2.5).

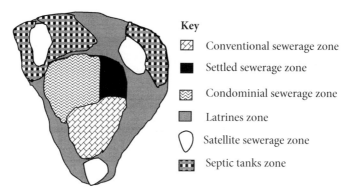

Figure 2.2. Illustration of mixed sanitary provision in cities with spatial variability.

2.4 Institutional dimensions of sanitary provision

2.4.1 Public utility provision

Sanitation services contribute to public health that is of such importance that it is provided as public and merit goods by public agencies. Rees (1998) notes that sanitary systems have inherent characteristics that make involvement of private companies not very likely, such as spatial monopolies, the public health and environmental benefits it brings, balancing affordability and subsidies, and its huge capital investments and sunk costs. Sanitary utilities, in fact, are considered natural monopolies, non-rival and non-excludable that can only be efficiently produced by government forces, with private sector either incapable to produce, and if they do, they are subsidised (Halcombe, 1997; McGranahan & Satterthwaite, 2006). Sanitary development is viewed also as a modernist project, where utilities are centrally planned, managed and regulated by the state through sectoral policies, standards, and development planning under state monopolies through standardised services that are accessible and ubiquitous to each household (Arts *et al.*, 2006; Spaargaren, 2003; Van Vliet, Chappells, & Shove, 2005).

Criticisms of the public utility paradigm have emerged and intensified over the last three decades, with calls for privatisation. There are those who view this call as a shift from simple modernity (modern period) to reflexive modernity (late/post-modern period) and those who view it as a weakening of the state and emergence of markets as alternative provision paradigm. A shift from simple to post-modernity is viewed to result in structures of modernity being increasingly questioned and weakened from increasing societal risks, complexities and the inability of the state to provide public goods and services through centralised planning, investment and control as it used to be the case in the modern period. Utilities, e.g. sewerage systems, consequently, are restructuring from integrated and centralised systems characteristic of modern period to splintered, fragmented and differentiated utility goods and services characteristic of post-modern period (Van Vliet, 2002; Van Vliet *et al.*, 2005). Restructuring can be viewed as a strategic retreat and renewal of the state through transferring some public duties to both society and market, and markets taking more public responsibilities (Arts *et al.*, 2006). Others view it as state failure depicted by absence of adequate institutional and political capacity (Khan, 2002), and thus promotion of the private sector is seen as more efficient in provision of utility services through a lens of neoclassical economic theory.

2.4.2 *Private utility provision*

In neoclassical economic theory, the role of government is viewed to be limited to ensuring markets operate freely, with services dictated by supply and demand, maximization of utility and profit, which in turn, drive production and consumption processes. Private companies are not new in water and sanitary services, with (Budds & McGranahan, 2003) noting that the first water and sanitation services were provided by the private sector to well-to-do communities who were able and willing to pay. Proponents of private provision view privately run utilities as cost-conscious, apolitical, and demand responsive (McGranahan & Satterthwaite, 2006). Since public authorities dominate utility provision, neoliberal proponents prescribed privatisation and private sector participation policies in an attempt to enable private sector access into public utility networks. They argue that privatisation would enhance efficiency, transfer financial burden from public to private investors, reduce poverty, and curb inequality in access to water and sanitary services (Castro, 2008).

Privatisation of public utilities, however, has had dismal impacts on urban sewerage provision, accounting for a very small market segment (Bayliss, 2003; Budds & McGranahan, 2003; Kariuki & Schartz, 2005; McGranahan & Satterthwaite, 2006). Private sector, moreover, has been selective and inclined towards large-scale networks that combine water and sewerage, large cities, large economies and large middle-class being 'cherry-picked'. This is in contrast to Sub-Saharan Africa where financing for sewerage utilities mostly come via the public sector and user charges (McGranahan & Satterthwaite, 2006; Budds & McGranahan, 2003; Gunatilake & Jose, 2008; Castro, 2008; Van Dijk, 2008). Private investments have benefited relatively wealthier countries, with for instance, Sub-Saharan Africa receiving 0.2% and other lowest income countries 1% of private investments for the period 1990-2005 (Budds & McGranahan, 2003; Gunatilake & Jose, 2008). Privatisation experiences from UK and Wales show that while current customers are being satisfied in terms of levels of service, there appears to be no incentive for long term investments, putting long term sustainability of asset management and serviceability into jeopardy (Ashley & Hopkins, 2002). Calls for private provision of water and sanitary services seems to be a mirage and a strategy to dismantle or reduce the public sector in delivery of essential public services, but it has failed as investments in water and sanitary infrastructures are still coming from the public sector and account for about 90% even during the height of privatisation (Castro, 2008). Having been the main proponents and drivers of private utility provision until recently, the World Bank has acknowledged that multinational private monopolies are neither investors nor developers, but profiteers (Bayliss, 2003; Castro, 2008). Therefore, revamping of public provision coupled with alternative provision arrangements to supplements each other is imperative. On the other hand, private sector contribution to the sanitation sector can bring in new market components and thus additional drivers which are not yet explored, such as valorising resources from the negative valued streams. Examples are minerals, nutrients (nitrates, phosphates, and potassium), energy, and stabilised matter.

2.4.3 Voluntary sector utility provision

The voluntary sector includes non-governmental organisations (NGOs), community based organisations (CBOs) and faith based organisations (FBOs) (Hasan, 1990, 2002; Picciotto, 1995; Gaye and Diallo, 1997; Krishna, 2003; Tukahirwa & Mol, 2010; Schwartz & Sanga, 2010). Picciotto (1995) notes that for effective market operation, voluntary organisations are needed in countries where the market and the state are poised in relative to fill gaps in provision, restrain the state, point out excesses of market, and provide avenues for participation and cooperation.

The proponents of voluntary sector perceive them as participatory, innovative, flexible, cheap and able to benefit the poorest of the poor, and they are alternative to public and private infrastructure and services provision in low-income areas that individuals cannot address by themselves (Stewart, 1997; Hasan, 1990). CBOs according to Hasan (1990) provide localised infrastructures, operate utilities and lobby government to improve infrastructure or services. The standard of infrastructures and services depend on skills and capacities in the community, varying from substandard and poorly maintained to excellent and well maintained. Such action is often self-financing and has no element of grant or subsidy coming from an outside agency. The main constraints facing these organisations are lack of finance, low personnel capacity, policy constraints, political interference, and their transitory or informal nature (Hasan, 1990; Tukahirwa, Mol, & Oosterveer, 2010).

Gaye and Diallo (1997) note that local sanitary problems can be solved by local communities in partnerships with NGOs, local authorities (LAs) and support from other agencies. This view is supported by Krishna (2003) who argues that the utility of both LAs and CBOs can be considerably enhanced when agencies work in partnerships with one another, with each playing different roles and responsibilities. Moreover, strong civil society strives in strong state, democratic and good governance and thus their entry in services provision does not mean weakening the state, but a strengthening and alteration of state intervention in delivery of services (Stewart, 1997).

2.4.4 Partnership utility provision

Public-private partnerships (PPPs) in its broadest sense entail public and private partners working together to achieve improved public infrastructure, community facilities and services (MMA, 1999; Weitz & Franceys, 2002; K'Akumu, 2006; UN-Habitat, 2003). Partnerships involve allocation of ownership, financing, and operation and maintenance responsibilities (Hukka & Katko, 2003). The partnership can be between household and public, community and public, public and private, private and private (Table 2.2). Partnerships are considered an alternative way of financing initiatives, which would otherwise not be realised because the best elements of the private, public and voluntary sector are combined and risks, benefits and responsibilities shared. Partnerships can bring a number of benefits such as improved quality of service, enhanced expertise, reduced political interventions in utility operations, expanded service coverage to more customers including the poor, and improved operational efficiency, management, and system performance (Van Dijk, 2006; MMA, 1999).

Some other forms of PPPs, apart from those in Table 2.2, are (MMA, 1999):
- Operate and maintain, where a private partner operate and maintain a publicly owned facility.
- Design-build, where a private partner is contracted to design and build a facility to the standards and performance requirements while the public takes over the ownership and operation of the facility.
- Turnkey operation, where the public provides the financing for the project but engages a private partner to design, construct and operate the facility for a specified period of time.
- Wrap around addition, where a private partner finances and constructs an addition to an existing public facility, operate the addition for a specified period of time.
- Lease-purchase, where the public contracts with the private partner to design, finance and build a public facility, then the private partner leases the facility to the public for a specified period after which ownership vests with the public.
- Built-own-transfer, where the private developer obtains exclusive franchise to finance, build, operate, maintain, manage and collect user fees for a fixed period to amortize investment, then reverts to the public.

Table 2.2. Forms of environmental infrastructure provision (Weitz & Franceys, 2002; K'Akumu, 2006).

Form of provision	Asset ownership	Operation & maintenance	Capital investment	Commercial risk	Duration (years)	Tariff regulation	Monitor quality
Household	private household	private household	private with public	private with public	indefinite	private	public
Community	community	community	public with community	public with community	indefinite	community	public
Small-scale providers	private business	private	private	private	variable	private	public
Public agencies	public	public	public	public	unlimited	public	public
Service contracts	public	private & public	public	public	1-2	public	public
Management contract	public	private	public	public	3-5	public	public
Lease contract	public	private	public	shared	8-15	public	public
Concession	public	private	private	private	25-30	public	public
Built-operate transfer	private & public	private	private	private	20-30	public	public
Divestures	private	private	private	private	indefinite	public	public

In addition to the mentioned contract types other forms exist, such as built-operate, built-own-operate and built-own-operate-transfer. Besides, others include franchise systems, revolving funds and micro-financing. Therefore, the range of PPPs possibilities are diverse and can be custom-made to suit local circumstances.

2.4.5 End-user participation

Participation has different forms and levels, which range from manipulation to collaboration and citizen control (Arnstein, 1969; Randolph, 2004). End-users can participate from decision making processes to implementation and monitoring and evaluation of policies and projects. Participation is contestable. On the one hand, it is argued that it leads to better access to services with aid of local experts, promote flexible and differentiated services delivery, secure long-term operation and maintenance sustainability, stimulate demand, ensure accountability, build consensus, make wise decisions, build local capacity and instigate technical and commercial innovations (Randolph, 2004; Murray & Ray, 2010; Jaglin, 2002; Odolon, 1998; Hegger & Van Vliet, 2010). On the other hand, it is argued that participation transfers costs from mandated utility agencies to low-income household, produces systems that are unstable, can create inequalities, and can lock disadvantaged urban settlements into sub-standard systems that are very difficult to upgrade (Jaglin, 2002).

Although participation has been criticized, it seems to be an alternative paradigm to technocratic approach to development processes, especially at household and community level. This is more so in increasing access to improved sanitary services, where the problem is attributed to failure of supply-driven and technocratic approaches to sanitary provision, which are expensive, do not meet household service demand and are heavily reliant on external support and solutions are not replicable (Jenkins & Sugden, 2006 as cited in Murray & Ray, 2010). Consequently, there is a shift from top-down, technocratic approaches that focus on monopolistic service providers and which reduces end-users to recipient of services, to participatory approaches that focus on waste producers and users, i.e. households or communities, as key stakeholders in sanitary provision (Murray & Ray, 2010; Mara, 2005). A participatory approach to sanitary is not restricted to decentralised sanitation, which is often at household or community level and viewed as simple, low tech, flexible and participatory solutions (Smith, 2005; Murray, Ray, & Nelson, 2009; Spaargaren *et al.*, 2005), but also apply in hitherto conventional urban systems, which are technocratic and monopolistic in nature. For instance, Nance and Ortolano (2007) found that good sewer performance was associated with community participation, especially in mobilization and decision-making phases and not so much in construction and maintenance. Participation even in onsite systems such as community participation in operation and maintenance of toilet blocks in Mumbai India has had mixed results (McFarlane, 2008 as cited in Murray & Ray, 2010). For any meaningful participation, it requires targeting potential end-users before the system is designed and tailoring sanitation schemes such that the outputs meet their specific needs in terms of location, quality, level, flexibility and state, besides better matching the local conditions (Murray & Ray, 2010; Jaglin, 2002).

2.5 Modernised sanitary mixtures as a flexible mix of technical and institutional dimensions

2.5.1 The MM approach

The MM approach has been postulated since 2005 as a modernisation strategy for environmental infrastructures and institutional arrangements (Spaargaren *et al.*, 2005; Van Vliet, 2006; Hegger, 2007; Scheinberg & Mol, 2010; Van Buuren, 2010; Scheinberg, 2011). The MM approach in this thesis, however, is utilised as a tool to conceptualise, assess and provide direction for improving sanitary infrastructures, eventually resulting in a modernised sanitary mixture (MSM). An MSM is achieved by organising all sanitary provisions in such a way that it results in a mix of scales, strategies, technologies, payment systems and decision making structures (Spaargaren *et al.*, 2005) that comply with specified sustainability criteria of public and environmental health, accessibility and flexibility. If implemented, such mixtures would lead to configurations that take the best features out of both conventional centralised, generally perceived modern, and alternative or traditional decentralised systems. This can be achieved by combining features of large-scale, high-tech and technocratic approaches, with small-scale, low-tech participative approaches into new forms in order to better fit the local conditions (Spaargaren *et al.*, 2005).

2.5.2 The MM dimensions

The dimensions of sanitary provision espoused by MM approach are four (Figure 2.3). The MM dimensions are:
- its scale, between large-scale, fixed price and small scale, flexible price systems. This thesis covers technical and spatial scale, large versus small;
- its scope of management between centralised monopolistic organisation and decentralised multiple providers;

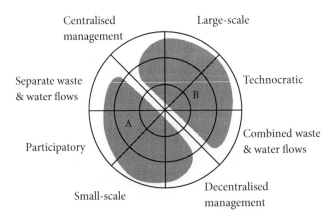

Figure 2.3. Modernised mixtures dimensions for classifying sanitary systems alternative to decentralised (A) and centralised (B) along multidimensional axes (Modified from Spaargaren et al.*, 2005).*

- the nature of flows, between separation and combination of water and waste flows;
- end-user participation, between participatory and technocratic (top-down) approaches.

2.5.3 Conceptual framework

The conceptual framework for this thesis (Figure 2.4) contextualises sanitary provision in East African cities in terms of spatial-technical and institutional dimensions depicted by four axes. Three concentric circles divide the axes into six scale numbers, which define service levels and are used to assess and map sanitary configurations as well. The resultant configurations can be framed conventional, traditional, and mixed or hybrid.

The sanitary options are then assessed on three sustainability criteria of public and environmental health, accessibility and flexibility. Sustainability assessment is imperative in determining whether existing sanitary systems can be judged as sustainable or not, with those unsustainable being a target for improvement or restructuring measures whereas those sustainable are replicated. A mix of technical and institutional arrangements is premised on the notion that merging them can lead to better sanitary provision. East African cities provide a good setting for the model since they are characterised by differentiated spatial structures, multimodal sanitary solutions and multiple providers, which goes beyond the centralised (conventional) and decentralised (alternative) approaches; and triad institutional pluralism models between public, private and voluntary sector.

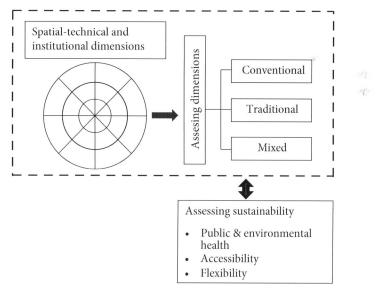

Figure 2.4. Conceptualising sanitary provision by dimensions, service levels, and sustainability criteria.

2.5.4 Assessment scales

The assessment scales within the MM approach applied in this thesis are:

a. Spatial-technical scale: large versus small scale systems

Scale can be defined in relation to the technical scale of implementation and coverage expressed in population size (P) and treatment capacity of STPs expressed in population equivalent (P.E.), whereas spatial scale relates to the area they occupy (ha) or the size of settlements expressed in population size (P). In Kampala and Kisumu, sanitation scales are not established. What exists is settlement size that is used to classify the size of urban centres based on population size. Merging land size and scale of sanitation systems is contestable. For instance, Van Buuren (2010) notes that maximum land area for a community and a small-scale sewerage are about 100 ha and 250 ha respectively. A neighbourhood, which is well defined since 1930s, is about 65 ha (Perry, 1939; Allaire, 1961). In this thesis, population size and spatial coverage are used to assess the MM dimension, large versus small scale (Table 2.3). The scale 4-6 is derived from Urban and Cities Act (Kenya, 2011).

b. Management arrangement: centralised versus decentralised

The management arrangement for service provision is often viewed as a triad: public, market and voluntary sector institutions, with partnerships in-between (Picciotto, 1995; Cohen & Paterson, 1999; Blair, 2001; Glasbergeren, Biermann, & Mol, 2007; Claassen, 2009; Tukahirwa, 2011). This thesis, introduces the fourth dimension, the household, as a service provider. Therefore, the triad institutional pluralism model is modified into tetragon to represent the provision reality in East African cities where the majority of households provide their own sanitary solutions (Figure 2.5). The scales for assessing the level of management are: (1) household, (2) community (NGOs, CBOs, FBOs, neighbourhood associations and cooperatives, (3) private commercial firms, (4) quasi-

Table 2.3. Assessment scales for large versus small scale sanitary provision.

Assessment scale	Spatial/service level	Population served (P)
1	• household • dwelling unit • housing cluster	5-50
2	• community	>50-1500
3	• neighbourhood	>1,500-5,000
4	• small urban	>5,000-50,000
5	• medium urban	>50,000-250,000
6	• large urban	>250,000

public institutions, e.g. universities, institutes, schools, hospitals, (5) semi-public authorities, e.g. local authorities and corporations, (6) state agencies, e.g. ministries, departments or directorates (Figure 2.5).

Mapping institutional arrangements along tetragon axis on six level scales (Figure 2.5) is viewed as establishing the relationship between sanitary systems and their management arrangements. The relationship can help in merging technical and spatial scale with required institutional arrangement through rebalancing of roles and responsibilities. Picciotto (1995) offers a way of rebalancing institutional arrangements through judicious mix of state, public, civil and market, and we add household, so as to achieve an appropriate balance and positive interplay between them. Rebalancing moves us away from one-model-fit-all to a MSM where provision is done by multiple service providers that are merged with scale of technology, nature of flows, and level of end-user participation against spatial structure.

c. Sanitary flows: separate versus combined waste and water flows

Separating storm water from sewage flow is attributed to the need for keeping toilet waste from diluting with large pool of water in order to reduce pathogenic risks. Wastewater flows can be separated based on source streams (Figure 2.6): urine, toilet flush, faeces, kitchen wastewater, anal cleansing, baths/showers, laundry, storm water and wastewater from industrial processes (Van Buuren, 2010). The rationale and incentive for separation is to recover nutrients, reuse wastewater, valorise sewage products, reduce treatment costs and apply appropriate technologies, but is yet to be proven (Van Lier & Lettinga, 1999; Kujawa & Zeeman, 2006; Zeeman, Kujawa, de Mes, Hernandez, de Graaff, Abu-Ghunmi *et al.*, 2008). Assessment for sanitary flows separation

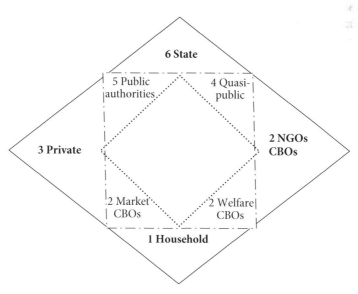

Figure 2.5. Assessment scales for management: centralised versus decentralised and public versus private provision.

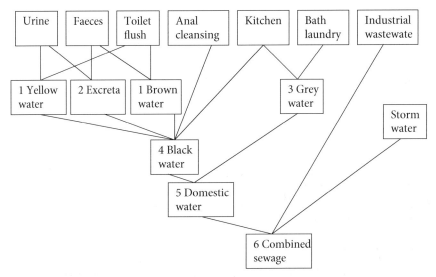

Figure 2.6. Sanitary flow assessment scales (modified from Van Buuren, 2010).

or combinations are based on source separation and concomitant reuse potentials: (1) urine separation, with or without flush water, i.e. yellow or brown water, (2) excreta, (3) grey water, (4) black water, (5) domestic, and (6) combined.

d. End-user participation: participatory versus technocratic

The assessment scales for the participatory-technocratic dimension are:
1. end-user participation in initiation, financing, construction and operation and maintenance;
2. end-user participation in initiation, financing and operation and maintenance, with construction by artisans;
3. end-user participation in resource mobilisation and selection of utility operators;
4. end-user participation in operation and maintenance;
5. end-user participation through awareness, sensitisation and satisfaction surveys;
6. no end user participation, except in payment of service charges and reporting complaints.

2.5.5 Mapping sanitary configurations

Upon assessing sanitary systems along four axes of MM dimensions and their respective six level scales as defined above, the resultant scales are mapped in the cells between the axes and concentric lines by way of shading. Suppose we have two systems to assess, X and Y. In system X, the urine is separated (scale 1 in flows dimension), is constructed by artisan but financed and initiated by households (scale 2 in participation dimension), located at household level (scale 1 in scale dimension) and managed by CBOs (scale 2 in management dimension). In systems Y, the system is managed by state (scale 6 in management dimension) is of medium urban in population (scale 5 in scale dimension), no end-user participation except in service charge payments (scale

6 in participation dimension) and domestic sewage is combined with industrial waste (scale 6 in flows dimension). These scales are then mapped in the cells between the axes and concentric lines by way of shading as shown in Figure 2.7 for system X and Y. Mapping sanitary configurations like this will help in a better understanding of them and provides a much clearer vision of how the various provision dimensions relate.

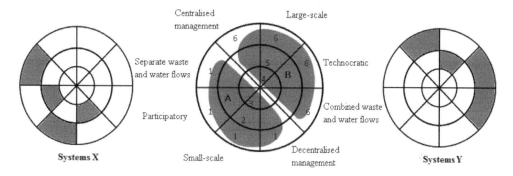

Figure 2.7. Mapping assessment scales by shading between axes and concentric lines.

2.6 Conclusion

Sanitary provision entails a plethora of scales, paradigms, flows and participation dimensions leading to sanitary mixtures. Sanitary mixtures, when framed on paradigms, can be conventional (modern), traditional (alternative) or mixed (hybrid). Institutional pluralism and multimodal sanitary systems call for rebalancing the existing socio-technical configurations so as to merge technical infrastructures with spatial structures and institutional arrangements. Such rebalancing calls firstly, for assessing sanitary dimensions in order to know which system can merge with which service provider, at what scale and spatial structure, and under which management and participation arrangement. Sanitary configurations can be better understood when mapped and schematically presented. Mapping is done by way of shading where scales from empirical analysis are transferred to axes within concentric lines that represent an assessment scale along MM dimension. Secondly, the sanitary systems should be subjected to sustainability assessment in order to determine which sanitary systems are sustainable, which are not sustainable and which elements need what kind of intervention measures.

Chapter 3.
Assessment of urban sewers and treatment facilities in Kampala and Kisumu as interplay of flows, networks and spaces

3.1 Introduction

The icon of urban sanitation modernisation in East African cities in the past was indisputably the construction of a centralised sewerage network complemented with 'Western' conventional treatment works consisting of mechanised aerobic treatment systems in an effort to modernise the townships and protect public health (Nilsson, 2006). Since 1980s, however, implementation of centralised systems in East African cities is viewed as unsustainable in the long run since their implementation are less successful, reinforces inequality, bleeds money out of social systems, and runs counter to local needs (Oosterveer & Spaargaren, 2010). The poor state of affairs is attributed to the socio-technical structures for water supply and sewerage that has remained largely unchanged since colonial time. This counts especially for institutional arrangments, standards and technological choices (Nilsson, 2006; Nilsson & Nyanchaga, 2008; K'Akumu, 2006), against dynamic and differentiated socio-spatial structure.

While significant steps have been made at the turn of the millenium to modernise technical and institutional arrangements of service provision, little is known about how they are reconfiguring sanitation practices and what structures are emerging. It is against this background, therefore, that the objective of this chapter is to assess and classify conventional urban sewerage systems as interplay of sanitary flows, networks and spaces, besides institutional arrangements. By examining such interplays, a set of criteria can be established on what areas to sewer, on the nature of sewer schemes to be used, on the kind of treatment works to adopt and on the nature of management arrangement that fit local conditions.

3.2 Approach and methodology

The findings are based on desktop studies, historical operation and maintenance records, technical reports[1], field surveys and in-depth expert interviews (Appendix 1). The research was carried out from October 2007 to December 2009.

Urban systems are assessed in two ways:
- First, assesment by the flows they convey (Section 3.3), the network they encompass (Section 3.4) and the spaces they exhibit (Section 3.5) to establish the feasibility for various sanitary solutions. To do this, water service level (l/ca*d or m³/d*ha), wastewater flows per day per hectare, i.e. base flow density (m³/d*ha), effluent discharge quality; extent and nature of

---

[1] Kampala Sanitation Strategy and Master Plan (NWSC, 2004), Kampala Sanitation Program Feasibility Study (NWSC, 2008), Kisumu Water Supply and Sanitation Feasibility Report (LVSWSB, 2005a), Long-Term Action: Plan: Sewerage Design Report (LVSWSB, 2008), Draft Practice Manual for Sewerage and Sanitation Services in Kenya (MWI, 2008b), and KIWASCO Water and Sanitation Sector Investment Plan (KIWASCO (2008).

sewerage network connections and mechanisation; sanitary space characteristics and population density (P/ha); and institutional arrangements are examined (Section 3.6). Wastewater flow estimates for sewerage are based on water consumption levels (l/ca*d or m³/d*ha), with the assumed sewage generation of 100 l/ca*d. Base flow density is begged on attaining sufficient concentration of medium and high income water users generating sufficient sewage flows for conventional gravity sewers to function sustainably. In Kampala city, the density for sewerage is set at 10 m³/d*ha and 200 P/ha (NWSC, 2004) whereas in Kisumu it is 120 P/ha (MWI, 2008b). However, for uniformity, base flow density is computed for Kisumu to be in tandem with Kampala. Water quality effluents were analysed using standard procedures (APHA, 992). Computations of STPs and land sizes (Table 3.2) are based on 260,920 inhabitants, emissions of 40 g BOD/ca*d, sewage flows of 26,092 m³/d, with applied design considerations from Metcalf and Eddy (2003) and Van Haandel and Lettinga (1994).

- Second, assessment along the four MM dimensions: scale, management, flows and end-user particpation (Figure 2.3). To do this, four multidimensional axes and six level scales (Chapter 2, Section 2.5.4) are used in the assessment. This is followed by mapping sanitary configurations by way of shading in the cells between the axes and concentric lines (Figure 2.7). Results are discussed in Section 3.7.

3.3 Sanitary flows

3.3.1 Generation, collection, treatment and disposal

Wastewater generation forecasts for 2010 in Kampala to sewers is about 25% of waterborne sanitation (septic tanks and sewers) (Figure 3.1). The catchments earmarked for sewerage are four

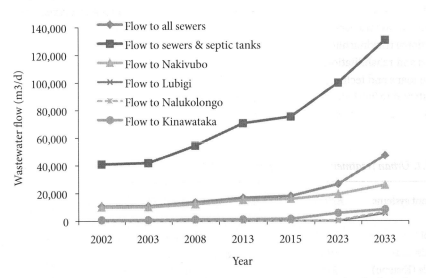

Figure 3.1. Wastewater flow forecast to waterborne sanitary systems in individual catchments and total sewers in Kampala (NWSC, 2004).

out of eight: Nakivubo, Lubigi, Nalukolongo and Kinawataka (Figure 3.4). The central Bugolobi STPs has a design load of 8,907 m³/d whereas satellite systems have combined capacity of 3,273 m³/d (NWSC, 2004). Existing sewerage systems in Kampala, therefore, can only convey and potentially treat about 23% of the flows. About 35% of the flow to Bugolobi STPs (Table 3.1) is attributed to water infiltration (NWSC, 2008), which is not included in the flows estimate.

The wastewater flow generation estimates for Kisumu in 2010 are about 26,000 m³/d (Figure 3.2) (LVSWSB, 2005a, 2008). Wastewater to sewers and septic tanks are approximately 21,000 m³/d. Based on design loads, existing STPs can treat 17,800 m³/d flows. However, the flows reaching STPs are 11,000 m³/d (Table 3.1) out of which only 9,800 m³/d can be treated, since Kisat design load has already been surpassed by 32%, whereas Nyalenda receive about 30% of its design load. Kisumu Molasses STPs is the only functional satellite system in Kisumu city, treating about 800 m³/d of industrial wastewater flow. Mamboleo slaughter house satellite STPs was abandoned in 1997, upon which generated sewage flow was connected to Kibos-Mamboleo sewer trunk line that feeds into Nyalenda STPs. The other industrial pre-treatment plants (Kisumu Cotton Mills and Kenya Breweries) are no longer in operation with the collapse of the factories. The projections show increasingly high flow of sewage to sewers in Kisumu between 2010 and 2030, which is attributed to the envisaged application of simplified sewerage in high density informal slum settlements (LVSWSB, 2005a, 2008). Slum settlements account for about 60% of city population (UN-Habitat, 2005), yet are dismally sewered.

The main treatment technology for Kampala central (Bugolobi) and Kisumu Central (Kisat) is conventional trickling filters, which treats wastewater from central city area. Both STPs are overloaded whereas Nyalenda pond-based STPs, which serve Kisumu Eastern catchment, is underutilised (Table 3.1). Half of Kisat stages are operational: 3 primary sedimentation tanks out of 6; 3 trickling filters out of 6; and 2 sludge pumps out of 4. The secondary sedimentations tanks are all operational. Nyalenda is one-third operational, with two pond series out of three filled with sludge whereas all facultative ponds (both operational and disfunctional) are completely covered with water hyacinth. Likely, hydraulic short-cuts further limit the potential capacity. The conventional trickling filters based STPs have at least collapsed once in the past. Bugolobi collapsed in 1980s and rehabilitation programmes have been undertaken between 2000 and 2008 through external loans and technical assistance. Kisat collapsed in 1990s and rehabilitation programmes, which started in 2007 are still ongoing through external loans and technical assistance. Thus, both Bugolobi and Kisat are partially operational, with a number of components inoperational, e.g.

Table 3.1. Urban treatment loads in Kampala and Kisumu (NWSC, 2004; LVSWSB, 2008).

Treatment systems	Existing load (m³/d)	Design load (m³/d)	Status
Bugolobi (Kampala)	12,000	8,907	overloaded
Kisat (Kisumu)	9,000	6,800	overloaded
Nyalenda (Kisumu)	3,000	11,000	underutilised

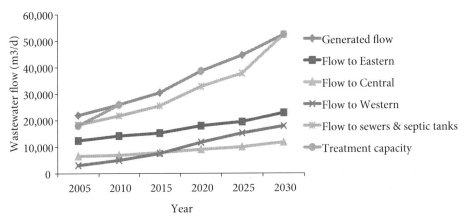

Figure 3.2. Wastewater flow forecast to sewers and catchments in Kisumu (LVSWSB, 2005a, 2008).

Bugolobi biogas plant, Kisat oil separator, 30% of trickling filters in Bugolobi and 50% in Kisat, and clogging of sludge digesters in Bugolobi.

Nyalenda was developed in 1976/77 to serve and take care of future wastewater flows from Kisumu Eastern catchment. The conventional trickling filters were developed in phases depending on sewer extensions and concomitant increase in sewage load:

- Phase I (1957): 1 primary sedimentation tank, 1 trickling filter, 1 secondary sedimentation tank, 1 sludge pump, effluent re-circulation pumping station, 2 sludge digesters, and 12 sludge drying beds with capacity of 1,500 m^3/d.
- Phase II (1961): 1 primary sedimentation tank, 1 trickling filter, 1 secondary sedimentation tank, and 1 sludge pump, resulting to 2,270 m^3/d capacity.
- Phase III (1966): 4 primary sedimentation tanks, 4 trickling filters, 2 sludge digesters and pumps, 2 secondary sedimentation tanks and 24 sludge drying beds, increasing treatment capacity to 6,800 m^3/d;
- Phase IV (1985/6): construction of oil separator due to high oil flow to the STPs.

The performance of treatment systems are poor (Figure 3.3), with effluents not complying with discharge standards (Uganda, 1999; Kenya, 2006). The stream to where Bugolobi discharges its effluent is polluted with sewage, with upstream and down stream of final Bugolobi STPs discharge point not meeting BOD, COD, TSS and faecal coliform discharge standards. Previous studies also show that the performance of treatment systems are not in tandem with discharge regulations for BOD, COD, TSS, P and N (LVEMP, 2001 cited in NWSC, 2004; JICA, 1998; LVSWSB, 2005a).

There is an apparent shift to hybrid treatment systems in Kampala and Kisumu through technology choice envisaged in the sewerage master plans. In the master plans, Kampala selected conventional trickling filters, upflow anaerobic sludge blanket (UASB) and waste stabilisation ponds, with post-treatment consisting of maturation ponds or natural wetlands (Table 3.2) (NWSC, 2004). In Kisumu, the choice is between conventional trickling filters and polishing with maturation ponds and grassplots or stabilsation ponds and polishing with grassplots (LVWSB, 2008). The selected treatment technologies are considered appropriate for the local conditions (Table 3.2)

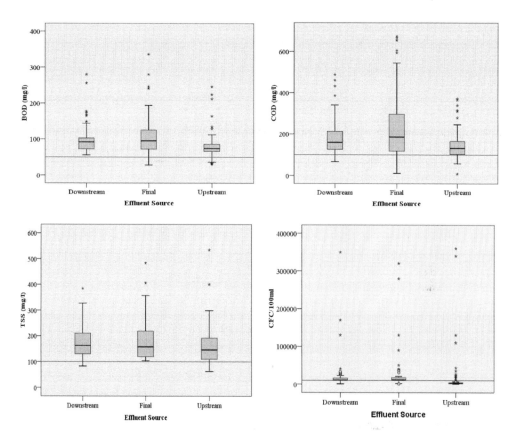

Figure 3.3. Box plots of overall effluent concentrations of selected water quality parameters of Bugolobi STPs in Kampala against upstream and downstream quality.
Symbol: Horizontal line indicate discharge standard for respective parameter

based on the level of instrumentations, skill requirements, and independence of foreign installation, investment and operation costs (NWSC, 2004). Another shift is the envisaged treatment of Nakivubo channel dry weather flow due to the high sewage content (Figure 3.3).

Recent data about stabilised bio-solids characteristics are lacking in Kampala and Kisumu. However, the LVEMP (2001) data used in Kampala sanitation master plan indicate that the final sludge from Bugolobi is not complying with generally acceptable standards for municipal sewage sludge reuse in agriculture with regard to viable intestinal nematode eggs and heavy metal concentrations (LVEMP, 2001 in NWSC, 2004) (Table 3.3). The high heavy metal concentrations are not attributed to domestic use alone, but from some industries located within the catchment of Bugolobi STPs. Despite the unacceptable quality of dried sludge, the sludge is sold for about €3.5 per tonne, and the destination and use of sludge is not known. There is reuse of stabilised bio-solids in Kisumu's Kisat, with major users being sugarcane farmers, but the quality is not known.

Nyalenda lacks anaerobic ponds, existing facultative ponds are not deslugeable, and drying beds. The biosolids are not recovered for reuse. The maturation ponds have fish stocks that the

Table 3.2. Treatment technology selection in Kampala (unshaded part is from NWSC, 2004; shaded part is computed).[1]

Technology process	Appropriateness[2]	Process option[3]	Land size (ha)	Remarks
Conventional trickling filters	++	PST-TF-W	21.2 (4.3)	selected
		PST-TF-M	19.4 (4.3)	
		PST-TF-M-W	29.2 (4.3)	
Anaerobic upflow sludge blanket (UASB)	+	UASB-F	22.0 (22.0)	selected
		UASB-F-M	27.3 (22.0)	
		UASB-F-M-W	44.2 (22.0)	
		UASB-F-W	33.9 (17.00	
Waste stabilisation ponds (WSPs)	++	A-F-M	39.8 (39.8)	selected
		F-M	68.9 (68.9)	
		A-F-W	39.8 (-)	
		F-W	49.0 (-)	
Wetlands (W)[4]	++	W/G	(-)	selected
Activated sludge	-	AS+SST	(0.39)[5]	unselected
Oxidation ditch	+	OD+SST	(1.42)[5]	unselected
Mechanical aerated lagoons	+	MAL	(2.52)[5]	unselected
Constructed wetlands	+	CW	(156)[5]	unselected
Rotating biological contactors	+	RBC+SST	(0.28)[5]	unselected

[1] *The design consideration used are 260,000 P.E., 40 g/P.E./d organic load, 26,000 dry weather flow, and 400 mg/l influent and 50 mg/l effluent BOD.*
[2] *-: low; +: average; ++: high.*
[3] *A: anaerobic; F: facultative; M: maturation; PST: primary sedimentation tanks; TF: trickling filter; SST: secondary sedimentation tanks; () land size based on organic removal.*
[4] *Tertiary treatment and nutrient removal.*
[5] *The unselected are computed for organic removal for comparison purposes.*

Table 3.3. Heavy metals in dried sludge in Kampala against Sewage Sludge Directive 86/278/EEC (NWSC, 2004).

	Pb (mg/kg SS)	Hq (mg/kg SS)	Cr_{total} (mg/kg SS)	Zn (mg/kg SS)	Cu (mg/kg SS)	Cd (mg/kg SS)	Ni (mg/kg SS)
Bugolobi	378	6.6	200	20,467	7,233	67	200
EU	750-1,200	16-25	1000	2,500-4,000	1000-1,750	20-40	300-400

surrounding community uses for domestic consumption and income generation, but the hygienic quality of the fish is not known. In addition, flows from maturation ponds are also used for irrigation of vegetables and tomatoes, with the farmers noting that treated wastewater contains nutrients and prevent pest. However, also the hygienic quality of the effluent is not known.

3.3.2 Determinants of sewerage development areas

This section examines the criteria that are utilised to establish sewerage areas in Kampala and Kisumu, water service level and population and baseflow density.

a. Water service level

In Kampala, introduction of piped water in 1927 was aimed at providing adequate water supply for non-African (European and Asian) population residing in Kampala township, Makerere college, Mulago hospital, the mission stations of Namirembe and Rubaga and Luzira prison in order to allow for future water-borne sewerage (Nilsson, 2006). Two water service levels were set, 180 l/ca*d for the non-Africans and 90 l/ca*d for Africans residing in Kampala. The 1930 sewerage design adopted British standards and practice such as house connection design, water consumption, and covered Kampala, Mulago and Makerere, areas that had high water consumption and the ability to pay (Nilsson, 2006). Sewerage development and extensions in 1950s, 1960s, and 1990s (Figure 3.5) followed the 1930 water supply and sewer policy. However, a shift is seen in the 2004 Sanitation Strategy and Master Plan and 2008 feasibility report, where reduced water service levels are set for sewerage planning: 150, 115 and 96 l/ca*d for high-income, medium-income and institutional residential respectively as potential sewerage areas (NWSC, 2004, 2008). Moreover, water consumption per hectare per day for sewerage connection is defined for commercial, industrial and non-residential institutions as 17.5, 15, and 10 m^3/d*ha respectively (NWSC, 2004). However, medium to low-income yard tap consumers with 16-40 l/ca*d and low-incomes stand pipes on 8-20 l/ca*d, as well as consumers supplied by alternative sources, e.g. springs, wells, boreholes and surface water channels, are not considered as potential sewerage areas. Public water supply covers about 71% of city population against 5% public sewerage (NWSC, 2007). Therefore, currently in Kampala, water supply is not a limiting factor for sewerage development.

In Kisumu, a sewerage scheme was proposed in 1928 after establishment of a piped water supply network in 1927. The scheme suggested a comprehensive sewerage for the whole of Kisumu Township with the exception of native locations and Old Kisumu, to be implemented in phases. First phase was to cover the high density and sanitary problematic middle township section, the second phase was to cover the low lying areas requiring lifting of sewage, and the third phase to cover the low density European and Goan residential settlements (Action, 1927). Although sewerage was developed in phases (1950s, 1960s, 1970s, and 1980s), it covered a whole township area where adequate water supply extended, but excluded African locations, thus it followed basically the 1928 water supply and sewer policy. In Kisumu, as generally in Kenya, a mix of British and American planning and service standards have been maintained since colonial time. During colonial time, water supply of 220 l/ca*d was adopted for sewerage in non-natives and 44 l/ca*d for unsewered Africans areas (Nilsson, 2008). More appropriate post-colonial standards were established in Kenya in 1973, which set water supply between 135-160 l/ca*d (WHO, 1972 in Nilsson & Nyanchaga, 2008). Practice Manual for Water Supply in Kenya sets different water consumption rates of 250, 150 and 75 l/ca*d for high-, medium-, and low-income respectively (MWI, 2005), compared to JICA (1998) standards as proposed for Kisumu of 200, 120 and 50-60 l/ca*d respectively, with the latter applicable to informal settlements utilising condominial sewerage. Moreover, the manual defines water supply threshold level for sewerage in industrial areas as 20 m³/ha*d. The 2008 sewerage plan for Kisumu does not include any area based threshold levels. Most Kisumu residents are not connected to piped water supply and get alternative water from public water kiosks, boreholes, springs and shallow wells, yet such sources are not recognised for sewer connections even if they are reticulated, e.g. Wandiege water supply and SANA project in Manyatta. Some water supply is executed through a delegated management model (DMM), but mechanisms of wastewater collection in line with this model is lacking. The regulation limits the amount of water supplied by private means to no more than 20 households, 25,000 l/d of water for domestic purposes or no more than 100,000 l/d for any purpose, except under the authority of a license (Kenya, 2002b).

The governmental policy in Kenya since independence is that urban systems for public water supply and sewage collection, treatment and disposal should be brought to and maintained in approximate balance (Kenya, 1974, 1994, 1997). The policy was refocused in 2007, with a target to reach through sustainable waterborne sewage collection, treatment and disposal 40% of the urban setting in 2015 and total sewerage coverage in all urban centres by 2030 (MWI, 2007). This is a shift in policy that is more in reality with urban structure which is differentiated (Figure 3.7; Table 3.7), which calls for different sanitary solutions and targets. In Kisumu, public water supply coverage is about 40% and concentrated in central business district, with extension to suburban developments and the airport. The 2030 planning horizon envisages 75% coverage (LVSWSB, 2008). Therefore, 100% sewerage coverage by 2030 is not tenable.

b. Population and base flow density

Parishes and sub-locations earmarked for sewerage in Kampala and Kisumu respectively are used to depict how population and base flow densities are used to determine sewerage areas and phase developments within and between sewerage areas.

Kampala City

Kampala's urban sewerage development trend is taking a catchment approach, considering the extensions in Nakivubo drainage catchment and proposed new sewerage areas in Lubigi, Nalukolongo and Kinawataka catchments (Figure 3.4; Table 3.4). Creation of Nakivubo catchment will necessitate extension of sewerage networks into surrounding parishes within the catchment based on base flow and population density. The parishes for sewerage extensions by 2013 are Bukesa, Namirembe, Kibuli, Kabagala, and Mengo (Table 3.4) (NWSC, 2004).

The proposed 2023 sewerage extensions are Kiswa, Katwe II, Wabigalo, Kisugu; and parts of Bukesa, Naguro 1 and Nakawa parishes (NWSC, 2004). Kinawataka sewerage catchment is scheduled for 2023 whereas Nalukolongo and Lubigi are scheduled for 2033 (Table 3.4; Figure 3.4).

Table 3.4. Population and base flow density as determinant of sewerage areas in Kampala.

Parish	Population 2033		Base flow density (m³/d*ha)			Area (ha)	Flow 2033 (m³/d)
	P/ha	Size (P)	2013	2023	2033		
Nakivubo sewerage catchment extensions							
Naguro 1	26.3	2,141	1.77	1.82	1.96	81	242
Nakawa	107.5	6,731	4.71	5.12	6.31	63	612
Bukesa	149.2	9,336	8.99	9.64[a]	11.23[b]	63	1,092
Nakasero 3	32.4	693	6.38	6.52	7.27	21	241
Kiswa	93.4	2,995	3.10	3.64	4.93	59	242
Namirembe	201.4[c]	5,992	*9.79*	**10.47**	**12.21**	30	562
Kibuli	241.3	8,306	*9.74*	*9.92*	**10.93**	34	582
Kabalagala	192.5	5,426	6.08	7.61	9.94	28	434
Mengo	201.4	4,059	*9.79*	**10.47**	**12.21**	20	381
Katwe II	263.2	8,312	7.04	7.76	**10.44**	32	510
Wabigalo	225.5	5,495	8.28	*9.15*	**10.73**	24	314
Kisugu	257.2	30,206	8.69	**10.90**	**14.25**	117	2,008
Bukasa	119.4	3,482	4.49	5.52	6.69	29	234
Lubigi sewerage catchment							
Bukoto I	187.3	32,774	5.33	7.80	**10.88**	175	2,285
Bwaise I	286.9	37,854	6.85	8.92	**11.11**	132	1,759
Kyebando	257.6	72,283	5.40	8.94	**14.02**	281	4,721
Makerere I	129.3	17,012	4.26	4.73	5.30	132	837
Makerere versity	69.2	4,238	4.76	5.76	6.39	61	539
Wandegeya	183.9	5,171	8.03	**10.89**	**13.89**	28	469
Nalukolongo sewerage catchment							
Kabowa	250.5	34,379	6.82	*9.20*	**12.10**	137	1,993
Kibuye II	203.7	4,720	8.81	*9.87*	**10.23**	23	284
Makindye II	190.9	7,688	6.00	8.62	**11.28**	40	545
Najjanankumbi I	213.5	12,206	6.04	8.62	**11.66**	57	800
Najjanankumbi II	134.1	8,525	4.77	6.33	8.02	64	612
Ndeeba	101.2	7,988	2.47	3.49	4.86	79	460
Kinawataka new sewerage catchment							
Mbuya I	257.6	12,030	7.87	**11.44**	**15.37**	47	861
Mbuya II	99.5	15,068	3.45	4.43	5.37	151	976
Mutungo	284.0	54,170	8.88	**11.05**	**14.83**	191	3,394
Ntinda I	54.7	18,482	7.97	7.73	8.22	338	3,333
Naguro I	-	-	1.77	1.82	1.96	-	-
Naguro II	-	-	8.40	**11.84**	**18.66**	-	-

[a] The numbers in **bold and italic** indicate the areas that are at verge of base flow threshold level.

[b] The numbers in **bold** indicate the areas that have attained/surpassed sewerage base flow density threshold. Base flow density threshold for sewerage is 10 m³/d*ha adopted from the master plan (NWSC, 2004).

[c] The numbers in italics indicate the areas that have attained requisite population density for sewerage. Population density threshold for sewerage is 200 P/ha adopted from the master plan (NWSC, 2004).

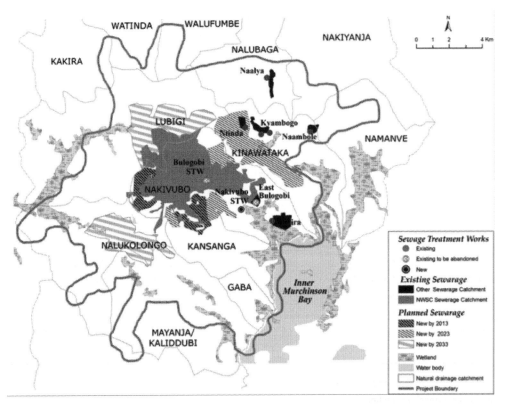

Figure 3.4. Kampala 2030 structure of sewerage type, coverage and catchments (NWSC, 2004).

Construction of Kinawataka will be in phases. The Trunk sewer, Mbuya I, Mutungo, Naguro and the STW is scheduled for 2023, whereas sewer lines for Ntinda I and Mbuya II are scheduled at a later stage depending on urban and demographic developments (NWSC, 2004).

Kisumu City

Further sewerage developments in Kisumu are geared towards increasing sustainability through cost-effective sewerage development and reduced operational and maintenance costs (JICA, 1998; LVSWSB, 2005a, 2008). This section examines population and base flow densities against sewerage development schedules across sub-locations in Kisumu[2] (Table 3.5) as a measure of sustainability.

Sewerage collection and treatment by the 2030 plan horizon is programmed to be carried out in three catchments, existing Eastern and Central catchments and a new Western catchment. Central catchment will be expanded from 385 ha to 437 ha and will cover mainly Kibuye and Milimani sub-locations (Table 3.5). Eastern catchment will be expanded from 214 ha to 358 ha and will cover mainly low-income areas of Nyalenda and Manyatta and parts of Milimani, Kibuye, Kasule and Wathorego (Table 3.5). The proposed Western catchment will cover 2,600 ha out of

---

[2] Figures are computed from population, water demand and wastewater projections from LVSWSB, 2008.

Table 3.5. Population and base flow density as determinant of sewerage areas in Kisumu.

Sub-location	Population 2030		Base flow density (m³/d*ha)				Area (ha)	Flow 2030 (m³/d)
	P/ha	Size (P)	2007	2010	2020	2030		
Kibuye	155.8[a]	106,733	9.84[b]	11.19[c]	14.11	18.22	685	12,483
Milimani	85.1	43,746	5.37	6.11	7.7	9.96	514	5,117
Kanyakwa	18.74	19,004	1.18	1.35	1.7	2.19	1014	2,222
Nyalenda	188	109,415	11.87	13.58	17.01	21.99	582	12,797
Manyatta	223.1	139,468	14.12	16.05	20.22	26.14	624	16,312
Wathorego	30.81	31,897	1.94	2.21	2.7	3.5	1035	3,618
Korondo	18.89	33,129	1.19	1.36	1.7	2.21	1754	3,874
Kogony	20.90	30,935	1.32	1.5	1.89	2.44	1480	3,618
Kasule	12.68	23,713	0.27	0.30	0.38	0.49	1871	925
Chiga	7.56	15,753	0.15	0.18	0.22	0.29	2083	614
Nyalunya	9.45	19,248	0.2	0.23	0.28	0.37	2035	750
Kodero	20.65	11,754	0.43	0.49	0.62	0.8	569	458
Got Nyabondo	17.77	14,980	0.37	0.43	0.54	0.61	843	599
Konya	19.73	22,842	0.42	0.47	0.59	0.77	1158	913

[a] The numbers in italic indicate the areas that have attained requisite population density for sewerage. Population density threshold for sewerage of 120 P/ha (MWI, 2008b) is applied in Kisumu.
[b] The numbers in **bold and italic** indicate the areas at verge of base flow threshold level.
[c] The numbers in **bold** indicate areas that have attained/surpassed threshold base flow density for sewerage. Kampala sanitation master plan threshold of 10 m³/d*ha for sewerage is used in Kisumu for comparison.

5,140 ha catchment area, which will collect sewage from Kanyakwar, Kogony; parts of Korando, Kibuye; and areas served by Mumias Road pumping station (Table 3.5). Besides, it will relieve pressure on Kisat STPs. Where it is technically and economically difficult to sewer, septic tanks are proposed for low-density areas. For those in low-income settlements, a mixture of public toilets with septic tanks and ventilated improved pit (VIP) latrines with a design that allows for faecal sludge emptying are proposed, accompanied by hygiene sensitisation campaigns.

3.4 Sanitary networks

The principal components of physical sewerage networks in Kampala and Kisumu are gravity sewers, inverted siphons and pumping stations. Sewerage schemes in Kampala and Kisumu cities are based on conventional sewers, which are typified by laying of pipes along the road and designed for at least once a day self-cleansing velocities, reduce build up of hydrogen sulphide and avoid abrasion caused by grit in the sewage. Where ground slope is steep and the flow velocity can cause abrasion, drop manholes are introduced to decrease the velocity (NWSC, 2004, 2008; LVSWSB, 2005a, 2008). Storm water is separated from sewage, which has been a design practice since 1930s. However, the infiltration rate in Kampala is about 35% whereas in Kisumu, large sections of sewerage areas have a low water table and are saturated throughout the year (NWSC, 2008; LVSWSB, 2005a, 2008). Consequently, infiltration is common in Kampala and Kisumu from groundwater arising from defective pipes or joints, service connections, and cross connections from storm water at uncovered manholes; the large areas they occupy, e.g. 214-1,800 ha and the large population they cover. Sewers are therefore designed for average peak wet weather flow and due consideration of infiltration. For instance, future sewers in Kampala are desinged to accomodate an infiltration of 20% besides the 1.5 peak factor for large population (NWSC, 2008).

The sewerage system in Kampala and Kisumu contains networks with characteristics displayed in Table 3.6. The sewerage network in Kampala comprises about 160 km of sewers with diameters ranging from DN 175-675 mm, with 64% of the pipes less than DN 200 mm and 22% between DN 200-300 mm.

The sewerage materials are varied, with asbestos, glazed vitrified clay and spun iron dominant in 1940s and 1950s while from late 1990s, concrete and PVC are increasingly being used (Figure 3.5).

Table 3.6. Sewer catchments, siphons and pumping stations in Kampala and Kisumu (NWSC, 2004, 2008; LVSWSB, 2008).

Sub-catchment	Type of system	Pump head (m)	Capacity (m³/h)	Siphon status	Penstocks	Area (ha)
Kampala city						
High level system						1,265[a]
• High level	gravity + siphon			partial	missing	350
• Kitante West	gravity + siphon			working		250
• Kitante East	gravity + siphon			not working	missing	240
• Lugogo valley	gravity + siphon			partial	partial	425
Low level system						735[a]
• Kibira Road	pumping	12.0	720			395
• Bugolobi	pumping	12.7	250			290
• East Bugolobi	pump + siphon	49.5	109	working	working	50
Kisumu city						
Central wastewater district						
• High level	gravity + siphon			partial	partial	385[a]
Low level system						
• Sunset	pumping	40	75.6			
• Kendu Lane	pumping	12	72			
• Mumias Road	pumping	10	97.2			
Eastern wastewater district						
• High level	gravity					214[a]

[a] Total catchment area.

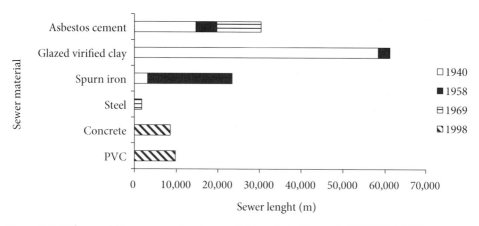

Figure 3.5. Urban public sewerage developmental trends in Kampala (NWSC, 2004).

The network is concentrated in the central part of the city covering some 2,000 ha. In Kampala, 37% of the sewage is being pumped and 63% transported by gravity and siphon arrangement. The siphons constructed in 1960s, Kitante East and part of low level, are not working; the Lugogo siphon is partially operational while penstocks in most siphons are either missing or not working properly (NWSC, 2008). The sewers are generally dug at a reasonable depth with an average depth of about 2 m. However, some 5% of the sewers are deeper than 4 m, with extreme depths of 8-15 m in Upper Kitante West, making safe access for maintenance at such depths very difficult. Some 5% of the sewers are laid with a cover of less than 1 m (NWSC, 2004).

There are two urban sewerage schemes in Kisumu, central and eastern catchments. The central sewerage network was constructed between 1955 and 1965 and expansion and maintenance programme undertaken between 1965-1975. It comprises of 9.5 km of sewers between DN 150-600 mm[3], 143 manholes[4], one siphon, and three operating pumping stations and two abandoned. The siphon comprises of pipes of size DN 225, 300 and 600. Only the latter is functional while DN 225 is leaking and the inlet valve was removed in DN 300. Eastern sewerage was constructed between 1975-1985, comprising of 8.5 km sewer line of size DN 175-675 mm, 298 manholes, and 0.3 m to 5.61 m sewer depths[5]. Some sections of sewers ranging from -0.4 m to -1.49 m are exposed and use steel pipes. Nyalenda STPs is meant to cater for the growth of Kisumu to the East, slum upgrading programmes in Nyalenda and Manyatta, and help eliminate the problematic Martin's Dyke and Nairobi Road pumping stations.

All pumping stations in Kampala and Kisumu have two pumps each, intended to operate on a duty and stand-by basis, with the exception of Kibira road with three pumps. The pumps are

---

[3] The minimum sewer material required is class 41 PVC pipe that should be non-absorbent and durable such that it can resist corrosion and abrasion (Kenya, 1999).

[4] Compared to Kisumu, the regulation stipulates that public sewers should have a nominal diameter of not less than 225 mm but local authority may approve a smaller diameter where appropriate (ibid).

[5] The depth of sewer pipes are determined by the appropriate fall in order to achieve self-cleansing velocity and adequately protected against damage due to external loads (Kenya, 1999).

operated manually because the automatic on/off switches are not working. Pumping stations are not operated at night due to security problems. As such, the night-flow is stored in the sewerage network, with risk of overflows through manholes when the system is full. In Kampala, the reported percentage average downtime of pumping stations for low level pump 1 (25%) and pump 2 (90%); Bugolobi pump 1 (100%) and pump 2 (20%); and East Bugolobi pump 1 (70%) and pump 2 (90%) (NWSC, 2004). In Kisumu, pumping stations and electromechanical equipment were reported to be grounded for over a decade (JICA, 1998; LVSWSB, 2005a), with rehabilitation programmes for the same undertaken from 2008 (LVSWB, 2008). The standby generators in Kampala's Bugolobi east and low-level pumping station have insufficient power to operate the pumps during power cuts. Pumping stations generally take a long time to be restored because spare parts purchases, which are often imported, take up to a year. Electricity bills and fuel costs are high, e.g. Kisumu electricty bills are often more than Kwh 7,500.

Manholes tend to be located at about 60-80 m[6] apart and in some cases, distances between manholes are shorter making maintenance more easy. In Kampala, approximately 10% and 40% of the manhole covers are missing and buried, respectively (NWSC, 2004). In Kisumu all iron-based lids have been stolen. Sewerage in Kisumu central wastewater district have 8% of manholes broken, 34% of covers are missing, and 22% have covers sealed (LVSWSB, 2008). Manholes covers are being replaced with concrete slab covers in Kampala and Kisumu.

The system of siphons consistsof 2 or 3 parallel pipes and in some cases duplicated chambers from which siphons convey the sewage flow to STW. Difficulties are experienced in the use of siphons: attainment of self-cleansing velocities and high tendency for blockages with subsequent overflow. They have to be isolated and drained, and blank flanges have to be unbolted before jetting (NWSC, 2004, 2008; LVSWSB, 2005a). Most of the siphons are located in built up areas and when blockages occur, sewage overflows along busy streets.

3.5 Sanitary spaces

Sanitary spaces entail land requirements for sewerage and treatment systems in order to ensure accessibility, compatibility, public health and environmental protection[7]. The basic requirements for connection and access to sewerage is that a customer should show proof of land ownership[8], a site plan showing the location of the plot in relation to adjacent plots[9] (Kenya, 1999), and a sewer line to be within 60 m (Kenya, 1986; Uganda, 2000). These requirements are achieved through planning policies that have bearing on urban sanition: designation, zoning and subdivision layouts.

---

[6] The regulation stipulates that manholes shall be provided in sewers at changes of direction, gradient and pipe sizes; at every junction; and at distances not exceeding 30 m for sewers with diameters not larger than 600 mm and 90 m for sewers with diameters larger than 600 mm (Kenya, 1999).

[7] Physical Planning Act (Kenya, 1996) prohibit or control the use and development of land and buildings in the interests of proper and orderly development of its area, the subdivision of land or existing plots into smaller areas, approval of all development applications, grant all development permissions, and ensure proper execution and implementation of approved physical development plans.

[8] Copy of title deed, land registration or plot number; lease certificate/search certificate or sale agreements or for tenants, tenancy agreement.

[9] From the Ministry of Lands or Town Planning Department, an approved spatial development plan.

Designation of townships began with gazettement of Kampala and Kisumu townships at the turn of 20th century. The Buganda agreement of 1900 divided Buganda Kingdom into Kampala administered by British colony and Kibuga administered by the King of Buganda (Nawangwe & Nuwagaba, 2002; Nilsson, 2006). Designation of Kisumu in 1903 led to immigrants administered by township authority (Board and later Municipality) while African areas were administrated by Local Native Councils (later African Development Councils). Designation of townships defined jurisdictional areas within which planning, control and sanitary service provision took place, and those excluded.

Zoning regulation during the colonial period was also used to exclude natives from township on sanitary and social grounds. For instance the 1908 Kisumu zoning led to partitioning of townships into three blocks, A, B and C (Figure 3.6) (UN-Habitat, 2005). Block A comprised the colonial and railway centres, and the European, Asian, Indian, and Railway residential areas. Block B was an undeveloped sanitary buffer between the immigrants (Block A) and the natives (Block C). Block C comprised the African native locations living in unplanned tribal villages. Segregation of Asians and European, however, was not considered necessary on sanitary grounds, but township authorities imposed discretionary sanitary, police, and building regulations without racial prejudice to achieve the same (Thompson, 1931). Planning was undertaken in colonial townships (Kisumu, 1908, 1936, 1960; Kampala, 1930, 1951) while African locations developed unregulated and unplanned.

The Buganda agreement (1900) and sanitary zoning structured later policies in Kampala and Kisumu respectively, creating duality between townships (immigrants) and native (African) areas. Duality is characterised by differences in administration, planning and service provision, resulting in sanitary divide thoughout colonial rule. Townships were administered by councils, with a mandate to provide municipal services, such as water supply, sanitation, and town planning through council departments and committees. Township councils charged property tax and service charge supplemented by financial allocations and loans from colonial government for provision of night soil and later sewerage through centralised bail and sewerage systems respectively. African areas were administered by native councils, Buganda Native Government for Kibuga and African

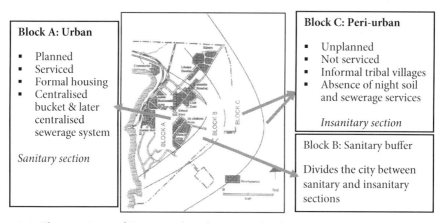

Figure 3.6. Characteristics of Kisumu colonial sanitary divide.

Native (Development) Council in Kisumu. Africans lived a semi-rural lifestyle, had no or low wages, and were not charged property tax or service charge. Native councils were mandated to provide sanitary services, but lacked organisational, financial and technical capacity, and were therefore unable to provide formal services. A sanitary divide, consequently, emerged between African areas and townships characterised by:

- designation of separate areas for immigrants (Kampala) and natives (Kibuga);
- physical divide as in Kisumu 1908 zoning that created a permanent sanitary buffer of 300 yards that literally divided township and African areas;
- service delivery divide where township councils are empowered to provide sanitary infrastructure and services while native councils were not;
- social divide where immigrants are segregated from natives in township through separate quarters for European, Asians, Indians and African government workers residing in townships;
- planning divide where townships are planned while African areas developed informally from rural tribal villages to overcrowded peri-urban settlements;
- separate councils for immigrants and natives, with the former supported by colonial budgetary support and later by none.

The sanitary divide created not only divide in terms of sanitation but also in broader modernity. The township was well planned and serviced with sewerage characteristics of a modern sanitary city while African areas remained unplanned and un-serviced relying on onsite sanitation (Figure 3.6).

After independence, integration of excluded African areas, which had developed as unplanned peri-urban settlements, started through boundary extensions and large-scale planning schemes. The extension was to reign in haphazard peri-urban settlements that housed a majority of township workers as well as requirement for more land for urban developments (Anyumba, 1995; Nawangwe & Nuwagaba, 2002). In Kampala, the boundary was extented in 1968, followed by integrative spatial planning in the Kampala Transport Master Plan (1968), Development Plan (1972), and Structure Plan (1994). Technically, the latter is still the official spatial plan despite it officially lapsed in 2004. The three plans were not implemented because they covered too large areas, seven times the size of the 1951 planning area, the political instability of 1970s and 1980s, lack of funds, and inability of the council to enforce development control on private, customary, and freehold land tenure. Consequently, Kampala expanded from 7 to 25 hills, resulting in development of unplanned and poorly serviced informal settlements that include some high-value houses occupied by the rich (Kaggwa, 1994). Kisumu boundary was extended in 1972. Kisumu's integration planning schemes include the Short-Term Development Plan of 1969 and the 1983 Structure Plan. Planning succeeded in public land in Kampala and Kisumu and in private land where large private land was acquired and subdivision layouts prepared before allocation. Private land accounts for 70% of land in Kampala (Nawangwe & Nuwagaba, 2002), while in Kisumu the 1972 boundary extension, which accounts for about 93% of Kisumu city, is 87.7% private (Anyumba, 1995). Sewerage covers the colonial urban boundary in Kampala and Kisumu, with some extensions to planned peri-urban developments in Kisumu. The low density residential areas targeted for sewerage in the 1930s, Kololo and Nakasero in Kampala and Milimani in Kisumu, are still served sustainably by septic tanks given their low population and baseflow densities (Tables 3.4 and 3.5). Planning control succeeded in the urban core, which is a public land, but failed in peri-urban areas, which is largely private land. The failure of large-scale spatial plans led to decentralised spatial planning (see Chapter 4), slum upgrading programmes, and multiple urban structures.

Slum upgrading policies were implemented in 1970s and early 1980s in Kisumu through World Bank 2nd Urban Project, but Kampala did not benefit because of political instability until 1990 when Namuwongo slum was upgraded. The upgraded informal slum settlements in Kisumu are Nyalenda, Manyatta and Obunga. Slum upgrading introduced differentiated standards and service levels, namely introduction of minimum water supply through water kiosks and yard taps as opposed to piped in-house supply and adoption of pit latrines as a mode of sanitation instead of public sewerage, bail system and private septic tanks. However, upgrading was only undertaken within the colonial township boundary. Sewers pass[10] through slums currently with dismal or no connection with continued uncontrolled development and informality.

Kampala spatial structure[11] is characterised by clustering of land uses and income classes (Figure 3.7). Kisumu spatial structure contains 42 km² urban and centrally planned areas, 53 km² peri-urban and largely informal slum settlements, and 202 km² rural areas that comprise a mixture

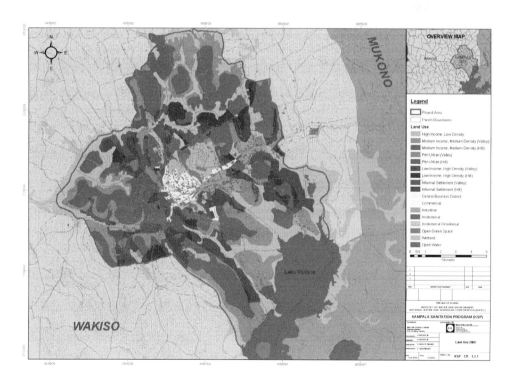

Figure 3.7. Kampala spatial structure (NWSC, 2008).

---

[10] Kenya (1999) stipulates a 3 m way leave be acquired where public sewer is constructed in a private land.

[11] Spatial structure can be categorised into urban, peri-urban, rural, slums, and suburban. Within each spatial structure, differences may exist in terms of planning, income, density, and service levels.

of unplanned, semi-controlled and suburban[12] settlements with significant portions on urban agriculture that came as a result of 1972 boundary extension. A multiple urban structure has shifted sanitation service level provision from one to multiple: conventional sewers, condominial sewers and a range of onsite options (Table 3.7). Settlements that are planned and based on medium-income (Kibuye & Kanyakwar) or high-income (Milimani); public land (Kanywakwar); or being located in designated industrial zones (Korando and Kogony) are target areas for conventional sewerage and full sewerage coverage irrespective of the density (Tables 3.5 and 3.7). High density informal settlements (i.e. Manyatta & Nyalenda) are target areas for a mixture of conventioal sewers, condominial sewers and onsite sanitation (JICA, 1998; LVSWSB, 2005a, 2008). Those targeted for conventional sewerage are settlements upgraded in 1970s and 1980s through slum upgrading programmes. The settlements outside designated sewerage areas (low density peri-urban and rural) were thought to continue being served by onsite systems.

Table 3.7. Kisumu urban structure and sewerage planning in eastern catchment (JICA, 1998).

Sub-location	Settlement structure	Income	Water (l/ca*d)	Sewerage service level	
				technology	Coverage (%)
Kibuye	urban	high	200	conventional	100%
		medium	120	conventional	100%
		low	60	condominial	100%
Milimani	urban	high	200	conventional	100%
		medium	120	conventional	100%
		low	60	condominial	100%
Nyalenda	peri-urban	high	120	conventional	100%
		medium	60	condominial	50%
		low	50	condominial	50%
Manyatta	peri-urban	high	120	conventional	100%
		medium	60	condominial	50%
		low	50	condominial	50%
Kasule	rural	high	120	conventional	100%
		medium	60	condominial	20%
		low	50	condominial	20%
Wathorego	rural	high	120	conventional	100%
		medium	60	condominial	80%
		low	50	condominial	80%

[12] Decentralised planned settlements located in Kisumu's peri-urban-rural interface e.g. Migosi, Kenya Re, Mamboleo, Kenya Ports Authority, Lake Basin Development Authority.

The space use requirement for buffer zones to avoid nuisance constitutes of a ring of preferably 500 m around treatment plants, away from any settlement, down wind from the community they serve, and away from any likely area for future expansion (LVSWSB, 2005a; Mara *et al.*, 1992). Earlier land use plans in Kampala and Kisumu located sewerage works at the edge of the master plan area to avoid nuisance. However, the city has expanded beyond the master plan area and around the treatment works, making treatment plants not complying with nuisance requirements.

3.6 Institutional arrangements

3.6.1 Organisational arrangement

The institutional framework in Kampala and Kisumu for sewerage is centralised at the national level and hierarchical in nature (Figure 3.8). Although the functions in Figure 3.8 point towards seperation of policy and coordination from regulation and asset ownership from service provision, the central government still controls the entire sector through the minister. The minister for the Minister of Water and Environment (MWE) in Uganda or Minister of Water and Irrigation

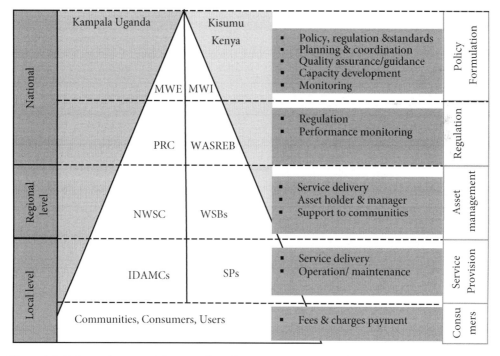

Figure 3.8. Roles and responsibilities in the urban sewerage in Uganda and Kenya.
Abbreviations: IDAMCs Internally Delegated Area Mangement Contracts; MWE Ministry of Water and Environment; MWI Minstry of Water and Irrigation; NWSC National Water and Sewerage Corporation; PRC Performance Review Committee; SPs Service Providers; WASREB Water Services regulatory Board; WSBs Water Servicers Boards.

(MWI) in Kenya, appoints board members to Water Services Regulatory Board (WASREB), Public Review Committee (PRC), NWSC, and Water Service Boards (SWBs) and gives it directions while the president appoints chairpersons to WASREB and NWSC. Consequently, decision making is centralised at the national level (K'Akumu, 2006; K'Akumu and Apida, 2006). NWSC and WSBs (LVSWSB) are public enterprises constituted as a corporation and responsible for asset holding and financing (Uganda, 1997a; Kenya, 2002b). Service providers (SPs) are Kampala Water Partnership (KWP) in Kampala and Kisumu Water and Sewerage Company (KIWASCO) in Kisumu. KWP is composed of NWSC employees who team up together to manage a service area as an operator. They are responsible for service provision through Internally Delegated Area Management Contract (IDAMC). KIWASCO is a public limited company whollyy owned by Municipal Council of Kisumu (MCK). Seperation of asset holding and service provision aims to commercialise service provision and attain financial sustainability. Private firms undertake design and construction under outsourcing contracts whereas communities are consumers and users of public services and their role is to pay service charges and meet contractual obligations.

3.6.2 Institutional reforms

In Kampala, urban sanitation has remained under public provision since 1900 to date. The privatisation wave in 1990s culminated in water sector reforms from mid 1990s (Table 3.8). It introduced New Public Management (NPM) arrangements (Uganda, 1995a,b, 1997a). A water and sewerage authority is appointed to operate an area through a performance contract (Uganda, 1995a, 1997a), which is administered by a corporation, partnership, trust, an individual or any other entity. The authority can enter into contract with any person or public authority for the provision of equipment, facilities, services or staff. Kampala is managed currently through a performance contract where the management fee is composed of base, performance and incentive fees, with a penalty for non-performance of 5%. The contract is multi-layered, e.g. government and NWSC; NWSC and KWP through IDAMCs; KWP and zonal managers through zonal performance contracts (ZPCs); KWP and third party for outsourcing service contracts.

In Kisumu, urban sanitation has remained under public provision since 1900 to date (Table 3.9). However, under structural adjustment programmes reforms tied to bilateral assistance in the 1990s, it led to a process of commercialisation in mid 1990s. A commission of Inquiry[13] on Local Authorities (LAs) in Kenya (1995) called for establishment of a separate corporation wholly owned by the council, but run on commercial basis. Urban Water and Sanitation Management (UWASAM) Project implemented from 1992 through 1999 led to establishment of Water and Sewerage Department from Engineering Department in 1993 and creation of KIWASCO[14] that is wholly owned by the council but fully responsible for their own financial affairs as the first step towards commercialisation. The Water Act (2002b) commercialised and liberalised water supply

---

[13] Via Gazette Notice Nos. 2939 and 2940 of 26th May 1995.

[14] Daily operations are run by a Board of Directors who appoints the Managing Director. The Board comprise of representatives from the council (Mayor, Town Clerk & Treasure), state (i.e. Ministry of Water & Local Government), and representatives (business community, women's organizations, & consumers).

Table 3.8. Trends of institutional reforms in sanitation sector in Kampala.

Period	Intervention	Features
1900	Gazetted as township	• mandate to provide sanitary services
1949	Kampala became a municipality	• made water and sewerage undertakers • Kibuga administered by Buganda Native Government
1968	Creation of Kampala City Council (KCC)	• Kibuga and Kampala municipalities were merged into KCC • KCC mandated to provide sanitary services in former Kibuga
1972	Presidential degree	• created National Water and Sewerage Corporation (NWSC) • Moved Water & Sewerage from council to NWSC
1995	NWSC Act	• commercialisation of NWSC services
1997	Corporate plans	• programmes planned through a 3 year corporate plans
1998	100 Days Program	• reverse the operational and financial inefficiencies
1999-2000	Service and Revenue Enhancement (SEREP)	• service enhancement through restoring customer confidence in NWSC operations through (SEREP I&II)
2002	Privatization of none-core services	• relinquish none-core services to service companies, i.e. guards, building maintenance and cleansing services
2002-2003	Stretch-Out Programme	• improve cash operation margins, reduce bureaucracy, introduce simplicity, and instilling self-confidence to workers
2002-2004	Management contract	• Ondeo Services Uganda limited (OSUL) awarded contract to rehabilitate network and modernise management systems, i.e. billing, network conditions, leak detection, and mapping
2003-2004	One-minute manager concept	• introduced individual accountability
2004	Tariff indexation	• annual tariff indexation adopted in Kampala
2004/06 2007/09	Kampala IDAMCs	• KWP was awarded to manage Kampala in order to promote autonomy and empowerment
2005-2006	Checker system	• enhance monitoring role of Head Office and improve efficiency in the operation of utilities
2004-2005	Zonal Performance Contracts (ZPCs)	• decentralisation of Kampala Area into zones • devolution of functions to the Zones where they assume operational functions, i.e. new connections, mains extension, leak control and billing
2006	Simplified Sewerage Connection Policy	• improve coverage and capacity utilisation • subsidised sewerage connection up to 60 metres from sewer line

Table 3.9. Trends of institutional reforms in Kisumu.

Period	Intervention	Features
1903	Township gazetted	• sanitary control and provision by the township board
1924	Local Native Ordinance	• was under District Commissioner and mandated to collect local taxes, and provide services, i.e. water and sanitation for natives
1928	District Council Ordinance	• segregated[1] councils: township Board for immigrants and Native Council for Africans
1945	Kisumu appointed water undertaker	• mandated to operate and maintain own water supply, sewerage and revenue collection • council sole provider of water and sewerage services
1950	African District Councils	• Local Native Councils changed into African District Councils
1954	By-Law	• lumped sanitary service charges
1974	Sewerage charges	• sewerage charges were introduced based on full cost recovery
1993	Creation of water and sewerage department	• delink water and sewerage services from Engineering Department • enhance efficiency and attention to water and sewerage services
1994	Separate account for water and sewerage services	• ring fence water and sewerage revenue from council general use • curb further deterioration of water and sewerage services • circumvent council bureaucratic procurements and employment process
1999	Formation of KIWASCO	• first step towards commercialisation of water and sewerage services • council as sole owner of the company owning 99.94% shares • Board of Directors runs day-to-day operations of the company
2001	Agency Agreement (AA)	• Kisumu Council and KIWASCO signed an agency agreement where KIWASCO will provide water and sewerage services in Kisumu on behalf of the Council
2003-2005	Agency Agreement	• commercialisation of water and sewerage services • separation of asset ownership, service provision and regulation
2005-2010	Service Provision Agreement (SPA)	• LVSWSB appointed water and sewerage licensee for Kisumu • five years lease of assets between LVSWSB and MCK • five year SPA between LVWSB and KIWASCO • SPA specify targets, performance benchmarks, tariff increase mechanisms, debt servicing and incentive fee structure

[1] Adopted as recommended by R.H. Feetham, Local Government Commissioner.

and sanitation in Kenya. Currently Kisumu is managed through a management contract, which sets performance standards, assessment procedures and indicators to be used as basis for rewards and penalties (LVSWSB, 2005b). Corporate business plans, which outline the strategic investment programme, customer focus and motivation packages, have been formulated to operationalize the performance contracts. The contract is multi-layered: development agreement between LVSWSB and Government; service provision license agreement between LVWSB and WASREB (regulator); a service provision agreement between LVSWSB and KIWASCO (LVSWSB, 2005b); a lease agreement between LVSWSB and MCK (LVSWSB, 2005c) and customer agreements between KIWASCO and customers.

3.6.3 Institutional models of sewerage provision

The institutional models for sewerage provision in Kampala are:
- Management contract
  Ondeo Services Uganda limited (OSUL) was awarded a 2 year management contract from 2002-2004 to rehabilitate the water and sewerage network and modernise management systems: billing, network conditions, leak detection and mapping. The management contract introduced performance based management, with incentives on performance and disincentives for misperformance. This type of management contract reflect a public-private partnerships, where NWSC is a public corporation and OSUL is a private company.
- Internally delegated area management contract (IDAMCs)
  The basic idea of IDAMCs is to encourage NWSC managers to form partnerships that take over the operating responsibilities in water and sewerage service area such as Kampala. IDAMCs aims at promoting autonomy and empowerment through seperation of the function of asset management from operations. In this contractual arrangement the operating partnership is acting as the agent and NWSC[15] head office as the principle. The key attribute of IDAMCs is that it offers an environment with strong incentives to perform; allows for guidance and support from the principle, prepare NWSC managers to operate as private operators in the future, and provide areas with greater managerial and financial autonomy. IDAMCs are litigation free partnerships, in which NWSC board is the final arbitrator in disputes and NWSC headquarters performs contract management and asset holding. Therefore, IDAMCs are public-public partnerships (Mutikanga, 2005).
- Zonal performnce contracts (ZPCs)
  During the year 2004/05, ZPCs were introduced to meet the increasing service demand through full decentralisation of services in Kampala area from KWP centred at Kampala 6th Street to 13 zones. Through ZPCs, all zones assumed responsbility for the operational functions such as new connections, mains extension, leak control, and billing. This turned zonal offices into one-stop centres for all services. ZPCs introduced competitive bidding in KWP through territorial management approach where KWP sets goals, performance targets, and standards to the zones. A score card is used to measure specific service levels achievement, which form

---

[15] The authority may enter into contract with any person or public authority for the provision of equipment, facilities, services or staff or joint use of the above (Water Act, 1997a).

the basis for employee rewards and penalties. At the end of each quarter, the results are made public through the NWSC Water Herald news letter and quarterly evaluation reports.
• Outsourcing
Planning and design of sewerage systems are tendered to private water and sewerage companies. Vehicle maintenance and cleansing services are also tendered out; so do rehabilitation programmes to private companies.

The institutional models for sewerage provision in Kisumu are:
• Asset lease agreement (ALA)
LVSWSB entered into ALA with Municipal Council of Kisumu (MCK) in June 2005 for a period of 5 years for lease of fixed assets, resolution of fixed liabilities, and resolution of customer deposits. This was preceded by inventory of fixed assets to be leased. Development of new fixed assets shall belong to LVSWB while capital improvements to existing fixed assets by LVSWSB shall be recorded in the inventory of fixed assets and jointly owned proportionately with the value of the capital improvements carried out. The Council recieves 5% of the collected revenue, which is reviewed regularly, with 0.5% penalty for failure per week.
• Services provision agreement (SPA)
KIWASCO provides services for a period of 5 years while LVSWSB acts as asset holder and responsible for capital investments. Any development agreement made by LVSWSB with the Government is supposed to be honoured by KIWASCO.
• Customer agreements (CA)
KIWASCO issue customer agreements to all customers within the service provider's area. Customers are to follow code of practice, which stipulates the terms and obligations:
  ○ ensure compliance with minimum technical standards;
  ○ ensure that bills for sewer service are paid even when there is no water connection;
  ○ monitor use of sewer line to ensure no hard objects get into the system in order to avoid sewer blockages;
  ○ make prompt payment of sewer bills by the 15th day of every month;
  ○ report all leaks and bursts promptly to ensure uninterrupted supply through calling hotline;
  ○ report all complaints and demand action from the company;
  ○ register to e-billing facility in order to make bill inquiry through cell phone.
• Outsourcing
Both KIWASCO and LVSWSB outsource feasibility studies, design, construction and rehabilitation works to private companies through open tenders.

3.7 Assessment of urban sewerage systems dimensions

Having presented the sanitary flows, networks, spaces and institutionl frameworks in Kampala and Kisumu, in this section, the urban sanitary system configurations are assessed against the MM dimensions (Figure 2.3) and six level asessment scales (Figure 2.7).

3.7.1 Technical and spatial scale: large versus small-scale systems

The scale of urban sanitary systems for Kampala central and Kisumu central and eastern catchment (Tables 3.4, 3.5 and 3.6) are medium-urban (Table 3.10; Figure 3.11).

Urban sanitary systems in Kampala and Kisumu are characterised by some major shifts in time. First, there is a shift from centralised to catchment approach to sewers and STW development (Figures 3.1, 3.2, 3.4; Tables 3.4, 3.5) in order to enhance operation and maintenance sustainability through minimal use of pumping stations, siphons and energy consumption. In Kisumu, estabishment of Eastern sewerage catchment disconnected from the Central catchment in 1977 led to abandonment of Martin's Dyke and Nairobi Road pumping stations, reduced operation and maintenance costs and eliminated overflow menace. Analysis show that a shift from a centralised system to four sewerage catchments in Kampala (NWSC, 2004) will result in:

- Lower overall expenditure on sewers and treatment over the next 30 years and beyond. Power costs due to sewage pumping will fall by 80% compared to current theoretical pumping costs, assuming all existing pumping stations operate as required.
- More lean and efficient personnel since no reductions in manpower is required, with the expanded sewerage network needing more personnel for maintenance.
- Sewerage will cover all commercial and administrative centre of the national capital, high-class residential areas, and industrial zones. Hence, its continuing functioning is of national strategic importance.
- More sustainable sewerage system with local environmental benefit since easier operation and improved capacity for future flows will help safeguard the environment due to reduced sewer flooding and in curbing partial discharge of treated sewage to Lake Victoria, which is also a source of water for cities and towns around it.

In Kisumu, the shift from existing two to three catchments will double the number of pumping stations from 3 to 6. This is because in Kisumu Western catchment, the suitable sites for STPs are already occupied whereas in and Eastern catchment, some settlements are lower than current Nyalenda STP, thus STPs located higher than some settlements, which require lifting of sewage to STPs by aid of pumping stations. The discounted net present value between collecting and treating sewage in existing two catchments or increasing to three is 2% (LVSWSB, 2005a). Therefore,

Table 3.10. Spatial-technical scales of urban sanitary systems in Kampala and Kisumu.

Spatial/service level	Population (P) served	Sanitary systems	Assessment scale
Medium-urban	50,000-250,000	Kampala Bugolobi central and proposed Lubigi, Nalukolongo and Kinawataka catchments Kisumu central and eastern and the proposed western catchments	5

further spatial decentralisation through catchment approach in Kisumu will not reduce operation and maitenance costs substantially. This would be different if pumping stations can be avoided through suitable approach, e.g. catchment or satellite such as in Kampala.

A second shift comprises the use of population and base flow density criteria to determine potential sustainability of sewerage areas. The average population densities in the planned sewerage areas in Kampala are Lubigi (209 P/ha), Nalukolongo (189 P/ha), Kinawataka (132 P/ha) (Table 3.4), in Kisumu catchment population densities are Central (344 P/ha), Eastern 544 P/ha) and Western (32 P/ha) (Table 3.5). Some parishes and sub-locations earmarked for sewerage (Tables 3.4 and 3.5) deviate from the set population and base flow density. The densities in Western catchment in Kisumu is low, but the biggest part of the catchment is an industrial area, e.g. Korando and Kogony sub-locations. The proposed 2023 Nakivubo sewerage extensions: Wabigalo and Katwe II meet density threholds in 2033; whereas Kiswa, Kisugu, Naguro I and Nakawa parishes do not meet the thresholds within the 2033 plan horizon. The densities in some parishes in Nakivubo extensions, Naguro 1; Nakawa and Nakasero 3; Lubigi (Makerere I); Nalalukolongo (Ndeeba); and Kinawataka (Mbuya II) will still be low enough for septic tanks to operate effectively beyond 2033 planning horizon. In Kisumu (Table 3.5), Kibuye, Nyalenda and Manyatta sub-locations attained required densities for sewerage by 2007 but are dismally sewered. Some sub-locations with very low densities, e.g. Milimani and Wathorego are sewered, whereas some with very low densities beyond the plan horizon are targeted for sewerage (Table 3.7). Nakasero and Makerere in Kampala and Milimani in Kisumu were targeted for sewerage in the 1930s, but septic tanks prevail in most parts of these areas to date. NWSC notices for mandatory connection to public sewer in Nakasero and Nakuro in 2004 failed. In Milimani, the part re-zoned for commercial use is sewered whereas the residential part is still on septic tanks. Although population and base flow density is a rational way for determining areas to sewer, however, their are some deviations. A number of factors can explain the deviations:

- Institutional land uses. Most of the flows from Kiswa, Katwee II, Wabigalo, Makerere University and Makerere I extensions will arise from institutional developments, which are often spatially concentrated, with large parts of the land left for recreational, circulation, demonstration functions or deferred for future expansion.
- Strategic and sensitive areas. Commercial areas, industrial zones, and government institutions, facilities and installations are targeted for sewerage. Such areas are strategic as they espouse public image of the state.
- Planned, middle and high income areas. Naguro 1, Nakawa, Nakasero and Kiswa parishes and Milimani and Kanyakwar sub-locations are easy to sewer and have the ability to pay.
- Supply driven service provision with cross-subsidisation. They require large contiguous areas to form economical sewerage service units with the assumption that the high density economical areas subsidises the low density uneconomical areas.
- Perception. Onsite sanitation is not often considered modern, thus always viewed as a transient stage towards sewer systems.

The third shift is the adoption of differentiated service levels depending on the area: conventional sewers, condominial sewers and onsite sanitation (Table 3.7). In planned areas, conventional sewers are applied where every plot ought to be individually connected to the public sewer. This is regulated by leasehold and building controls, e.g. plot ratio, line setback and architectural and structural plans. Conventional sewers are supported by planning schemes that require sanitary line layouts (Figure 3.9b), where each plot or premise is individually accessible by road. Large part of the city, however, consists of unplanned settlements with insanitary layouts (Figure 3.9c) that are inadequately served by roads (Nawangwe & Nuwagaba, 2002; UN-Habitat, 2005, 2008). Insanitary layouts defy spatial planning policies such as master and structure plans (Kombe, 2005;Olima, 1994; Nawangwe & Nuwagaba, 2002) and conventional sewerage requirement that cities be engineered into pipe-like entities (Nielsen & Clauson-Kaas, 1980; Newman, 2001; Graham & Marvin, 2001). Unplanned areas require acquisition of 3 m way leaves (Kenya, 1999; NWSC, 2004, 2008), but it can be done without way leave acquisition and population relocation if condominial sewers as proposed for Manyatta slum settlement in Kisumu coupled with emptying of chambers at accessible locations. JICA (1998) projected that adoption of condominial sewers in Kisumu would lead to 75% sewerage coverage by 2015. Sanitary lane layouts[16] (Figure 3.9a) can support condominial sewerage and are provided for in the regulation, e.g. prohibition of back to back dwelling and requirement for rear access of building from streets of not less than 1.6 m and a foot path of not less than 1 m in width (Kenya, 1999).

3.7.2 Management arrangement: centralised versus decentralised

Examining the institutional arrangements in Kampala and Kisumu, urban sanitary systems are provided by public agencies under new public management (NPM) arrangements (Table 3.11; Figure 3.10).The assessment of sanitation management arrangements on a scale of 1 to 6 on the management dimension (Figures 3.10 and 3.11), shows that the configurations in Kampala and Kisumu are ranked at 5 and 6. This is because development, operation and maintenance are to some extent decentralised or delegated to organisations or private firms at lower levels than central city level. However, financing, regulation and monitoringare still highly centralised. Hence, the ranking of 5 and 6 in the configuration schemes of Kampala Central, Kisume Central and Kisumu Eastern.

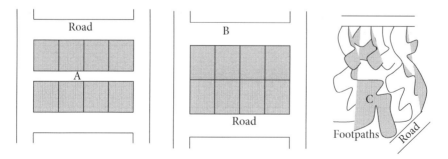

Figure 3.9. Subdivision layouts: sanitary lane (A), sanitary line (B) and insanitary line (C).

---

[16] Bucket sanitation system layout that still exists in old colonial neighbourhoods, civic and business areas.

Table 3.11. New public management aspects in Kampala and Kisumu.

NPM aspects	Kampala	Kisumu
Marketization	• outsourcing of services through tendering, e.g. planning and design, construction, vehicle maintenance, rehabilitation of STPs and cleansing	• outsourcing of services through tendering, e.g. sewerage design and construction, vehicle maintenance and rehabilitation of STPs
Decentralisation	• NWSC decentralised operation to KWP • spatial decentralisation of Kampala area into 13 zones	• LVSWSB delegated provision to KIWASCO • zoning of service areas into 3 zones
Efficiency	• performance contracts – IDAMCs/ZPCs • approved 3 year business plan • performance appraisal • incentive system	• performance contracts – SPA & ALA • approved 5 year business plan • performance audits • e-billing
Accountability	• creation of customer care division • publishing performance results in Water Herald Newsletter • published and uploading annual reports	• customer satisfaction surveys • customer care response policy

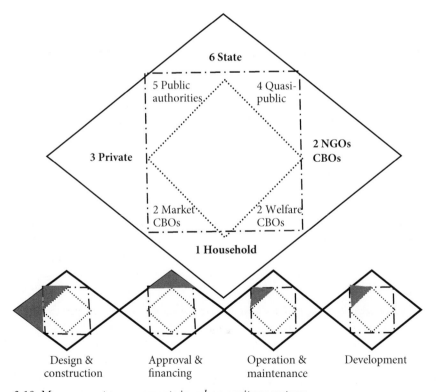

Figure 3.10. Management arrangements in urban sanitary systems.

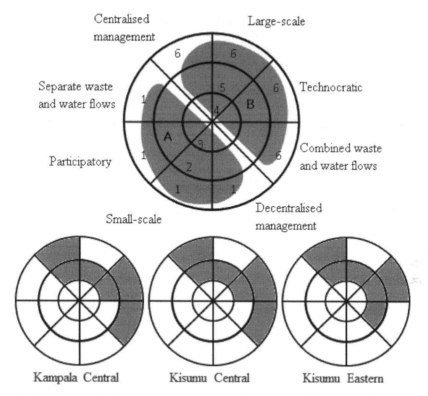

Figure 3.11. Assessment of urban sanitary systems configurations in Kampala and Kisumu against 4 MM dimensions on 6 point scale. Areas (A) and (B) represent decentralised and centralised paradigms respectively.

Examining the roles, responsibilities and practices (Figure 3.8; Tables 3.8 and 3.9; Section 3.6.2) in Kampala and Kisumu cities, the mode of urban sanitary provision is through public enterprises (Table 3.12) who operate with performance contracts with the government. Separation of policy and coordination from regulation is weak in Kampala compared to Kisumu. Services provision is separated from asset ownership, with the former through 3 year contracts in Kampala and 5 in Kisumu. Asset holders are responsible for new infrastructure development and major maintenance, whereas the operator is responsible for operation, billing, revenue collection, and limited maintenance. KWP receives base operational cost from time to time from NWSC and deposits revenue collected to NWSC controlled accouts. KIWASCO controls its accounts and manages its budgets. Service areas are spatially decentralised through creation of zones with zonal managers operating on constracts (ZPCs) in Kampala and none in Kisumu.

In Kampala, both IDAMCs and ZPCs are centrally monitored and controlled through checkers system. Assets in Kisumu are jointly owned by MCK and LVSWSB in proportion to the rehabilitated assets and new developments in line with the 2005 inventory and resolution of asset agreements. In Kampala and Kisumu, outsourcing to private firms through open tender service contracts is a common practice. In both Kampala and Kisumu, households manage sewerage

Table 3.12. Service provision models in urban sanitary systems in Kampala and Kisumu.[1]

City	Form of provision	Assest ownership	Operation	Major maintenance	Duration (year)	Tariff regulations	Quality monitoring
Kampala	MC	NWSC	OSUL	NWSC	2	MWE	MWE
	IDAMCs	NWSC	KWP	NWSC	3	MWE	NWSC
	ZPCs	NWSC	KWP	NWSC	1/4	-	NWSC
	outsource	NWSC	KWP	NWSC	variable	tendering	NWSC
Kisumu	AA	MCK	KIWASCO	KIWASCO	indefinite	MWI	-
	SPA	LVSWSB	KIWASCO	LVSWSB	5	WASREB	WASREB
	ALA	MKC	LVSWSB	LVSWSB	5	fee	inventory
	CA	LVSWSB	KIWASCO	KIWASCO	indefinite	WASREB	WASREB
	outsource	LVSWSB	KIWASCO	LVSWSB	variable	tendering	LVSWSB

[1] MC: Management contract; IDAMCs: Internally Delegated Area Management Contracts; ZPCs: Zonal Performance Contracts; AA: Agency Agreements; SPA: Service provision agreement; ALA: Asset lease agreement; CA: Customer agreements; NWSC: National Water and Sewerage Corporation; MCK: Muncipala Council of Kisumu; LVSWSB: Lake Victoria South Water Services Baord; KIWASCO: Kisumu Water and Sewerage Company; MWE: Ministry of Water and Environment; MWI: Ministry of Water and Irrigation; WASREB: Water Service Regulatory Board.

infrastructures within their properties, upto inspection chambers, but with supervision by experts. Public management principles have been embraced in Kampala and kisumu (Table 3.11), which is exhibited by martketisation, decentralisation, efficiency and accountability.

3.7.3 End-user particpation: participatory versus technocratic

In urban systems in Kampala and Kisumu, householdsor community groups as end-users are not involved, except by participating in surveys, e.g. for willingness to pay, ability to pay, satisfaction surveys and in sensitisation programmes. Rules and responsibilities for end-users are well defined, but mostly comprise of obligations of end-users towards connections, payment of bills and reporting sewage overflows and blockages. Consequently, the assement scale for end-user particpation dimension for Kampala and Kisumu urban sanitary systems is at 5 and 6 (Figure 3.11).

3.7.4 Sanitary flows: combined versus separate water and waste flows

Wastewater generation sources in Kisumu Eastern catchment are mainly domestic from residential and institutional land uses. Wastewater generation sources in Kisumu and Kampala central catchments are mainly combined: domestic, industrial and storm water. The latter is due to cross connections via inspection chambers, faulty manholes and infiltration. Thus, following the modernised mixture assessment on the MM dimension regarding sanitary, Kisumu Eastern catchment is assessed at 5 and Kampala and Kisumu Central areassessed at 6 (Figure 3.11). Reuse and resource recovery practices are through sale of bio-solids for agricultural use at Kampala and Kisumu central STPs and irrigation of crops using wastewater from or fishing in Nyalenda maturation ponds. El-Shafai et al. (2004) showed that fish cultivated in sewage treatment ponds may be contaminated with pathogenic bacteria. At Nyalenda STPs, the hygienic quality of the fish effluent is not known. Treatment systems also receive septage from onsite systems, which is co-treated with sewage.

Effluent discharge standards are stringent, sometimes contradictory and very difficult to achieve[17]. For instance, the nitrogen removal is condradictory since the standard for NH_4^+-N is equal to TN and NO_3^-. The requirement for phosphorous removal is not achievable without additional expensive technologies that may not be sustainable considering the level of economy of East African states. The incorporation of P and N in discharge standards requires advanced wastewater treatment technologies or land based treatment with a high area demand. The latter then requires a long conveyance system to cheap land areas. Both approaches demand considerable financial resources. Starting with a basic primary treatment and partial secondary treatment for all citizens seems to be more cost-effective and would address about 80% of the polution load. The standards in Kenya, especially for E. coli and coliforms, are very strict and seemingly unrealistic and are therefore generally ignored in the design of treatment works. Instead, the Ministry of Water an Irrigation (MWI, 2005) standards, which are adopted from WHO, are applied. Regulations in Uganda provide three options for disinfection of sewage to meet coliform discharge standards: chlorination, UV radiation and sewage treatment ponds (Uganda, 1999). Each of these alternatives have operational limitations (NWSC, 2004):

- Chlorination has high running costs, poses health risks for operating staff and potential threat to aquatic animals due to toxic by-products when chlorine reacts with organic matter. The latter is highly plausible since many treatment systems are overloaded and organic free effluents cannot be guaranteed.
- UV radiation is expensive and heavily dependent on imported equipment both for construction and spare parts.
- WSP require lots of land and thus expensive conveyance systems, but are cheap to operate and work efficiently under favourable climatic conditions in East Africa. Operational problems are related to efficiency loss owing to sludge accumulation, short circuiting and overloading.

Apparently, the adopted treatment technologies cannot meet the regulatory requirement, especially for ehavy metals (Table 3.3), demanding more advanced treatment systems (Table 3.2) (NWSC, 2004; LVSWSB, 2005a, 2008). The planned use of maturation ponds, natural wetlands, and grassplots to post-treat STPs effluents follow guidelines to comply with faecal coliform requirements and nutrient removal. The preference from the traditional use of conventional trickling filters to pond

---

[17] Standards for effluent into the environment are: (a) Kenya (2006); BOD (30 mg/l), COD (50 mg/l), TSS (50 mg/l), E. coli counts (Nil/100 ml), total coliform counts (30/100 ml), and detergents (Nil/mg/l); (b) Uganda (1999); BOD (50 mg/l), TSS/COD (100 mg/l), TN/NH_4^+-N/$N0_3$ (10 mg/l), NO_2 (2 mg/l), TP (5 mg/l), Ortho-PO_4 (5 mg/l) and FC (10,000/100 ml); (c) MWI (2005); BOD (50 mg/l), COD (100 mg/l), TSS (100 mg/l), E. coli counts (1000/100 ml).

systems translates into substantial (cheap) land requirements which are not readily available in and nearby cities (Table 3.2). In addition, the required conveyance systems can be very expensive whereas current discharge requirements are becoming more stringent. Limited land availability, increasing effluent restrictions, increasing energy concern, and recognition of the values of wastewater bound resources, seems to drive the shift from a restricted technological approach, i.e. a choice between conventional trickling filters or pond systems, to a more open technology search, better fitting the local socio-economic conditions.

3.8 Conclusions

The configurations of urban sanitary systems (Figure 3.11) are medium-urban in scale, technocratic in approach andmainly of combined sewage flows, with dismal reuse and recovery practices. The current approach to urban sanitary provision is convential and modernist in approach. Consequently, it is expensive to operate and maintain, and defy existing socio-spatial-technical structures, but a shift to catchment approach to enhance its reflexiveness and sustainability was observed. Besides, stringent environmental standards, energy concerns, and ongoing recognition of wastewater-bound resources such as energy, nutrients, water and stabilised matter has seen some shifts towards hybrid treatment systems. Despite institutional reforms in the sector over the last two decades, urban sanitary systems are still centrally planned and managed, but some new public management principles have been introduced such as outsourcing of planning and design services, decentralisation, performance contracting and accountability.

Reconsidering urban sewerage and treatment systems as interplay of sanitary flows, networks and spaces helps in defining and designing differential service levels: technology choice, scales, and standards, resulting in socio-technical mixtures as espoused by MM approach. Regonition of such interplay may help in planning and designing new infrastructures and modes of governance for providing sanitation for all.

Chapter 4.
Potentials of satellite sanitary systems in Kampala City

4.1 Introduction

In most large towns of developing countries, public sewerage coverage is dismal and this situation is likely to remain since governmental agencies in most of these countries are not capable of providing other infrastructural utilities than the way they have been doing in the past (Lee & Floris, 2003). Sewerage provision in East African cities are done through government-driven centralised planning, financing and management based on sewerage master plans. This provision arrangement, however, has failed to keep pace with rapid urban growth leading to alternative sewerage delivery independent of urban public sewerage systems. A potentially feasible shift to alternative sanitary provision in East African cities is to establish decentralised satellite sewerage and treatment systems. In East African cities of Nairobi (Kenya), Dar es Salaam (Tanzania) and Kampala (Uganda) there are over 30 satellite sewerage systems, which serve residential settlements, industrial complexes, campuses, schools, airports, stadium and disciplined forces facilities, e.g. military barracks, prisons and police stations (AWSB, 2005; DAWASA, 2008; NWSC, 2004). A satellite approach to sanitary provision is not a new phenomenon since satellite sewer systems have been in place since the 1940s in Nairobi and 1960s in Kampala (AWSB, 2005; NWSC, 2004); and they are increasing in number (Figure 4.1).

In satellite areas, decentralised spatial planning, sewerage, treatment, and drainage are converging to provide small-scale solutions aimed at curbing urban informality, spread of insanitary conditions and environmental pollution. The satellite approach is a shift from conventional provision and combines the promotion of multiple systems (Ho, 2005; Van Dijk, 2008; Massoud, Tarhini, & Nasr, 2009; Gikas & Tchobanoglous, 2009) with a decentralised approach to water management ascribed to multi-centred city development, where small-scale semi-collective systems can be realised in communities without substantial piping (Newman, 2001). As East

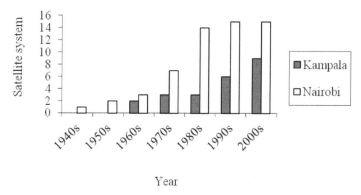

Figure 4.1. Number of satellite systems over the decades in Kampala and Nairobi.

Africa is not on track to meeting the MDG target (WHO/UNICEF, 2010), any improved sanitation, especially alternative sewerage development that complements public provision, is welcomed.

Public sewerage in Kampala is lagging behind population growth (Figure 4.2) and city coverage (Figure 3.4). Kampala Sanitation Strategy and Master Plan (NWSC, 2004) envisages 25% of both population and land coverage by 2033 (Figure 4.2). The envisaged sewerage development and expansion in the master plan, except Bugolobi and Ndinda, will not reach the existing satellite areas by 2033 (Figure 3.4). Eventually it has taken public sewerage 80 years to cover 5% of the population and 10% of the city area, while it still followed colonial spatial and sewerage planning legacy 40 years after independence. The inability of public sewerage to keep pace with population and spatial growth provide opportunities for satellite systems outside public provisioning to develop.

Satellite systems have expanded from 0.8% of population coverage in 2003 (NWSC, 2004) to 1.4% in 2010, thus accounting for about 20% of the sewered area. The satellite approach is emerging as an alternative sewerage provision pathway not only in Kampala, but also in other East African cities, although precise and up to date data are hardly available. Satellite systems have the potential to adequately address sanitary provision in specific parts of East African cities, outside the centralised public sewerage areas. The aim of this chapter, therefore, is to describe satellite systems characteristics in Kampala, assess their performances and map their configurations along the MM dimensions.

4.2 Methodology

This chapter is based on case studies of both existing and planned satellite sewerage systems in Kampala. Asessement Data were derived from interviews with randomly selected satellite owners and operators, residents living close or being served by satellite systems, a Kampala city planner,

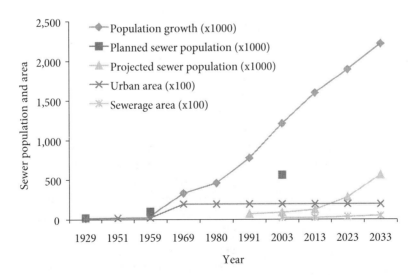

Figure 4.2. Public sewerage growth and population coverage in Kampala.

a sewerage consultancy firm and public health officers. Besides, structured observations coupled with documents and records were made.

Satellite systems are assessed in four ways:

- First, assessment of satellite settlements, sewers and STPs characteritics through interviews, visual inspection, site enquiries, analysis of policy, legal and regulatory documents and technical reports.

- Second, assessment of satellite STPs loading rates based on assumed sewage production of 100 l/ca*d, per capita BOD discharge of 40 g/ca*d, loading rates – anaerobic 500-10,000 kg BOD/ha*d, facultative 150-500 kg BOD/ha*d, maturation 50-150 kg BOD/ha*d. Facultative pond loads of 300 BOD/ha*d are used given Kampala temperature range of 20-25 °C. Formulas used in calculations are:

$$Q = \text{Specific flow} \left(\frac{1}{\text{ca*d}}\right) \times \text{Population (ca)} \tag{1}$$

$$\text{HRT} = \frac{V\ (m^2)}{Q\left(\frac{m^2}{d}\right)} \tag{2}$$

$$\text{Surface BOD loading} = \frac{\text{BOD discharge} \left(\frac{g}{\text{ca*d}}\right) \times \text{Population (ca)}}{\text{Surface area (ha)}} \tag{3}$$

$$\text{Volumetric BOD loading} = \frac{\text{BOD discharge} \left(\frac{g}{\text{ca*d}}\right) \times \text{Population (ca)}}{\text{Volume } (m^3)} \tag{4}$$

$$\text{Current capacity use (\%)} = \frac{\text{Current surface BOD loading}}{\text{Design surface BOD loading}} \times 100 \tag{5}$$

- Third, assessments[18] of STPs water qality performance by utilising seven grab samples for TSS, COD, N-NH$_4$, and TP and five grab samples for faecal coliform (CFU). Water quality assesments performed for this research, however, are limited to five treatment systems: Bugolobi, Naalya, Namboole, Ntinda and Unise. The five cases are chosen because they are accessible, have operated for long, and most of their treatment stages are operational. Those left out are Luzira, which is within a maximum security prison and thus inaccessible; Kyambogo, which is only partially operational because the second pond is disused; Mukono because it had not operated for long; and Naguru, which is no longer operational, with sewerage dilapidated and treatment ponds disused for decades. Upon design of the monitoring programme, sampling and analysis was done by National Water and Sewerage Coroporation (NWSC) as part of Lake Victoria Environment Management Programme (LVEMP). In Kampala, the programme monitors water quality along Nakivubo-Inner-Murchison Bay in Lake Victoria. Effluents were analysed using standard procedures (APHA, 1992).

- Fourth, assessment of satellite systems configurations along four MM dimensions: scale, management, flows and end-user particpation. To do this, four multidimensional axes (Figure

---

[18] February 8, 15, 20, 22 and 27; July 17; and December 5, 2008.

2.3) and six level scales (Chapter 2, Section 2.5.4) are used in the assesment. This is followed by mapping sanitary configurations by way of shading in the cells between the axes and concentric lines (Figure 2.7). Results are discussed in Section 4.7.

4.3 Institutional arrangements for sewerage systems

Satellite systems are acknowledged in the current Sanitation Strategy and Master Plan (NWSC, 2004) as 'other sewerage catchments' (Figure 3.4). Deliberate attempts have been made to stimulate private sector efforts in the development of affordable, planned and serviced settlements to bridge a housing deficit that is estimated at 100,000 units[19]. A satellite approach offers a better impetus for such initiatives. The regulation (Uganda, 1995a, 1995b, 1997a, 1997b, 1999, 2000), which assigns duties and responsibilities to various actors in the sector (Table 4.1), provides for private sewer developments in accordance with code and standards of workmanship. The codes and standards relate to matters of design and construction protocols, type of materials, fittings and appliances, and implementation of works by duly qualified and authorised persons, under supervision of and with approval by the sewerage authority.

The STPs are supposed to be developed by public sewerage authorities, e.g. NWSC for Kampala, whereas private sewers are expected to connect to public sewers and handed over the asset ownership rights and management to the area sewerage authority. The Public Health Act (Uganda, 2000), which KCCA enforces, requires provision of adequate sanitation by a developer: sewerage, septic tank or any other approved sanitation. The Water Statute (Uganda, 1995a) provides for development of community water and sanitation facilities. DWD approves sewerage and treatment plants, whether public or private. NEMA regulations (Uganda, 1999) stipulate that treatment works should be approved, registered, operated with an annual waste discharge permit and monitor effluent compliance. The Water Statute and Water Act (Uganda, 1995a, 1997a) provides for appointment of a sewerage authority to operate and maintain a sewerage area in all cases through a declaration or performance contract, with clearly set service levels, targets and compliance rules. They also provide for levying of approved sewerage tariffs to meet operation and maintenance costs.

Considering the institutional arrangements available (Table 4.1), sewerage is thought of as a public provision. Private provision is limited to development of sewers under supervision of authorised persons and sewerage authorities and connection of such sewers to public sewers. A city authority area is still considered a sewerage service area, yet a number of sewerage catchments (Figures 3.4 and 4.3) exist within the same city area, which are operated by different entities (Tables 4.8 and 7.2). However, the ambiguities in regulation provide a leeway for development of decentralised satellite sewers and treatment works. For instance, whereas the Water Statute and the Water Act gives NWSC mandate as sole sewerage authority in Kampala, the Public Health Act and NEMA regulations as well as approval by Directorate of Water Development (DWD) do not discriminate whether provision is by NWSC or by private parties.

---

[19] Through reduced value added tax by Ministry of Finance in 2007/08 National Budget on sale of residential properties from 18% to 5%.

Table 4.1. Sewerage and treatment system management arrangements in Kampala city.[1]

Function	DWD	NEMA	NWSC	KCCA	Developers	Consultants	Explanation of management arrangements
Development control				X			• development control through zoning, subdivision, change of user, spatial planning • approvals of subdivision, architectural and structural plans • awarding certificate of occupancy
Sewerage planning	X		X			X	• undertaking feasibility studies, planning, design and construction
Sewerage provision			X		X		• NWSC provide sewerage services as the mandated sewerage authority for Kampala • construction of private sewers and their connection to public sewers or treatment works • handing over of privately developed sewers to area sewerage authority
Sewerage approval	X			X			• approval of sewerage and treatment systems
Sewerage operation and maintenance			X				• appointment of sewerage authority on basis of performance contracts, investment plans and approved tariffs
Network extensions			X				• reticulations expansion through connection fees
Wastewater treatment			X				• operation of works based on annual discharge permit charged on volumetric BOD loads • registration of works • compliance to discharge standards
Storm water drainage				X			• separate storm water and sewage drainage • prohibits discharge of sewage to storm water drains
Discharge permits	X	X					• operation of treatment systems with a discharge permit from NEMA or on delegation by DWD
Annual monitoring	X	X					• monitoring by NEMA or on delegation by DWD to ensure sewerage authorities meet set targets and comply with standards

[1] DWD: Directorate of Water Development; KCCA: Kampala Capital City Authority; NEMA: National Environment Management Authority.

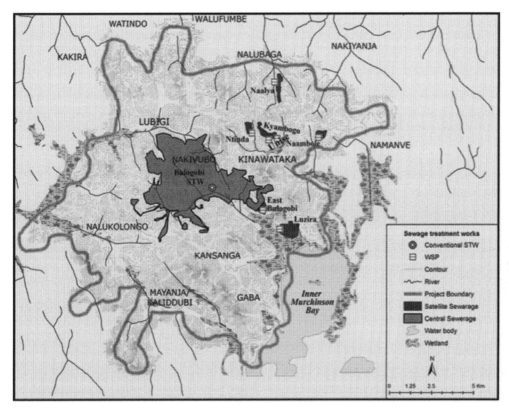

Figure 4.3. Urban and satellite sewerage and treatment areas in Kampala (NWSC, 2004).

4.4 Satellite settlement characteristics

Satellite settlements in Kampala are a product of blocks of land that are decentrally planned and serviced with localised sewerage and treatment systems (Tables 4.2 and 4.3), connected to public water supply and the national electricity grid. Satellite areas meet planning criteria required by KCCA that plots be surveyed in blocks for planning purposes (Nkurunziza, 2007). They are located outside the public sewerage area (Figure 4.3) and colonial township area. They were developed during the 1960s for government facilities, during the 1970s for government residential schemes and increasingly from 1990s for private residential and institutional settlements (Table 4.2). The settlement size comprise of about 1,500-14,000 P, a density of about 30-190 P/ha, and an area of about 20-330 ha (Table 4.2). Satellite areas are land use specific serving residential settlements, higher educational campuses, a stadium and police and prison facilities. They are developed as part of projects, e.g. housing or government facilities, but not as independent infrastructure development. The settlements are closed and exclusionary, e.g. developed with a particular population, area and target group. The treatment systems are located at the edge of settlements with less than a 50 m buffer strip.

Table 4.2. Socio-economic characteristics of satellite settlements in Kampala.[1]

Satellite area	Size (ha)	Year built	Developer	Density (P/ha)	Income status	Sewerage status	No. of units
Bugolobi	24	1970	NHCC	186	middle	existing	986
Kyambogo	331	1964	University	N/A	endowed	existing	N/A
Kiwatule	20	2009	NHCC	75	middle	existing	268
Lubowa 80	17	planned	NHCC	47	high	proposed	150
Luzira	-	-	Government	(-)	subsidised	existing	N/A
Mukono	-	2007	University	N/A	endowed	existing	N/A
Unise	90	1964	Government	N/A	endowed	existing	N/A
Naalya	42	1998	NHCC	48	middle	existing	320
Naguru	-	-	Government	(-)	subsidised	disused	-
Namboole	-	1997	Government	N/A	endowed	existing	N/A
Namungoona	150	2009	NHCC	93	middle/high	proposed	2,368
Nansana	24	planned	Arkwright	75	middle	proposed	300
Ntinda	44	1993	NHCC	34	middle	existing	94
Royal Palms	150	planned	Night Frank	53	middle/high	proposed	1,300

[1] N/A: Not Applicable; -: Data unavailable; NHCC: National Housing and Construction Company; P: population.

Satellite settlements are developed for middle and high income groups, endowed institutions and government facilities (Table 4.2). The middle income residential areas are Ntinda, Naalya and Bugolobi, whereas the high incomes are Royal Palms and Lubowa. The endowed institutions are Mukono and Kyambogo universities; whereas the government facilities are Mandela National Stadium, Luzira Prison and Uganda National Institute for Special Education (Unise). The residential housing is developed mainly from loans from local commercial banks, with land and property used as collateral. The housing units are bought in one instalment through mortgage or loan schemes from commercial banks, with developers recouping their investments upon sale of properties while banks deal with repayment from customers. The costs of sewerage and treatment infrastructures are factored in the property development lumped with other utilities and costs are recouped when properties are sold. Endowed institutions develop their premises from a number of ways, e.g. Kyambogo, a public university, receives government allocation supplemented by their own revenue generation; Mukono, a private university, is financed through loans and own revenue generation; whereas Namboole, Luzira and Unise are financed through continued government support (Table 4.2). Upon sale of residential properties to individual buyers, mechanisms for transfer of ownership of developed sanitary infrastructures are lacking.

Satellite projects have high political support, as ministers or even the president have launched or commissioned them. Developers in satellite settlements (Table 4.2) include government (stadium, prison and Naguru), quasi-public institutions (Kyambogo and Unise), private institutions (Mukono

University), a public housing company (NHCC), and private housing companies (Night Frank and Arkwright). Moreover, the National Social Security Fund (NSSF) intends to enter the satellite market with even larger areas and population equivalents. For instance, some of the proposed housing estates such as Nsimbe (340 ha and 5,000 units), Lubowa (226 ha and 3,000 units), and Temangalo (184 ha and 5,000 units), will result in P size of about 15,000-30,000.

Given the foregoing settlements characteristics, a satellite approach to sanitation provision may have a number of advantages:

- Can be part of decentralised sewerage and treatment system in peri-urban and rural areas where public sewerage cannot be connected cost-effectively.
- Does not rely on external loans, grants and subsidies, which are controlled by government, thus amenable for domestic resource mobilisation.
- Targets a small population, area and land use, thus have small footprint, is demand driven, and tailor made to the needs of specific user groups.
- Complies with planning and infrastructure standards as it needs approval before seeking investment capital. Kampala is largely an informal area where development organically grows outside the centrally planned city without control. Thus satellite settlements make a contribution in curbing the spread of haphazard and insanitary developments in pockets of Kampala.
- Has access to multiple financial providers, tapping resources from public, private, quasi-public and state entities.
- Satellite sewerage is based on gravity only within a catchment and functions independently from pumping stations and siphons, which increases operational robustness and reliability.

4.5 Performance of satellite systems

So far satellite sewer characteristics, their location, size and materials, have been inadequately documented. Interviews with sewerage managers, field observation and examination of technical reports revealed that satellite sewerage: (a) are conventionally planned, designed and constructed to comply with sewer standards and constructional protocols; (b) the sewer materials are mainly concrete for the systems built in the 1960s and the 1970s and PVC for the systems built from the 1990s on; and (c) sewer pipes in 5 of the 8 existing satellite systems are generally in good working conditions with no reported cases of sewer collapse, permanent blockages or continuous overflow. Naguru sewerage network (meant to serve Naguru Police Headquarters and Naguru Hill) is dilapidated and in a state of disrepair. Inlet pipes to Bugolobi and Kyambogo treatment ponds are broken such that sewage overflows before it enters the treatment ponds. The broken pipes have not been repaired for a decade. Satellite sewers and treatment systems are designed for a fixed population size without room for further connections from adjacent properties.

Satellite treatment systems discharge their treated effluents to rivers and wetlands (Figure 4.3). The systems are mainly waste stabilisations ponds with an activated sludge (AS) in Mukono, and a moving bed biological reactor (MBBR) procured for Lubowa 80 housing. The treatment systems consist of 2 to 4 treatment ponds, laid out in series and designed for 6-42 d retention time (Table 4.3). Most ponds, except Naalya, are not provided with a screen, a flume or by-pass control valves, and therefore cannot permit desludging of individual ponds while operation is being maintained in the others. The by-pass control valves in Naalya treatment ponds, however, have

Table 4.3. Assessment of Kampala satellite treatment systems.

Treatment Pond[1]	Bugolobi		Kyambogo		Unise			Naalya				Namboole			Ntinda		
Design pond type	F	F	F	F	F	M	M	F	F	M	M	F	F	M	F	M	M
Pond area (m²)	2,501	1,681	3,725	1,064	800	600	600	4,042	4,005	2,666	2,624	1,885	644	672	3,174	2,904	3,216
Pond depth (m)	1.21	1.21	1.25	1.25	1.25	1.30	1.14	1.96	1.54	1.4	1.37	1.72	1.39	1.05	1.25	1.3	0.89
Design population	1,050		2,300		720			4,600				1,550			3,200		
Design flow (m³/d)	105	105	230	230	72	72	72	460	460	460	460	155	155	155	320	320	320
Design HRT (d)	24.8	16.1	17.6	4.5	10.3	7.7	7.0	14.1	11.4	6.8	6.6	15.9	4.1	3.5	10.8	10.1	8.1
Design loading (kg BOD/ha·d)	167.9	157.1	247.0	112.0	360.0	88.3	40.8	455.2	119.6	81.0	54.0	328.9	156.0	110.0	403.3	98.2	25.6
Current population	4,500		6,320		783			2,000				1,550			1,500		
Current flow (m³/d)	450	450	632	632	78.3	78.3	78.3	200	200	200	200	155	155	155	150	150	150
Current HRT (d)	6.7	3.8	7.4	1.7	9.5	7.1	6.5	32.4	26.3	15.7	15.2	15.9	4.1	3.5	23.0	21.6	17.3
Current loading (kg BOD/ha·d)	719.7	673.5	678.7	307.7	391.5	96.0	44.3	197.9	52.0	35.2	23.5	328.9	156.0	110.0	189.0	46.0	12.0
Pond removal efficiency (%)	37		87		82	54		74	55	34		84	26		78	71	
Current/design loading (area-based)	4.3	4.3	2.8	2.8	1.1	1.1	1.1	0.4	0.4	0.4	0.4	1	1	1	0.5	0.5	0.5
Current pond type	A	A	A	A	F	M	M	F	F	M	M	F	F	M	F	M	M

[1] A: anaerobic pond; F: facultative pond; M: maturation pond.

rusted completely. Most ponds are partially covered by duckweed, water lettuce and embankments vegetation; Bugolobi and Unise are mostly affected. Naalya's last two ponds have fish stock that personnel fish for subsistence or sell for extra income.

In Kyambogo, the inter-pond pipe that connects facultative to maturation pond is completely blocked and has not been repaired for a decade. Naguru pond does not exist anymore as the ponds have been completely disused for decades. The site is covered by shrubs and a section is being used for agricultural purposes. The AS system at Mukono, which serves Uganda Christian University, has five functional units: equalization, aeration, sludge holding, clarifier, and disinfection. The plant is designed to treat 320 m³/d of sewage, but currently receives about 135 m³/d of sewage, with sludge wasted for one hour daily. The daily energy requirement of the AS plant is about 48 kWh/d based on the power rating of the backup generator in place. The proposed Lubowa 80 housing MBBR plant comprises of a buffer (holding) tank of 1,020 m³ and two-step MBBR with clarifier, coupled with sludge pump, blower, hydro cyclone and has an energy consumption of 53 kWh/d based on generator power rating.

Assessments of satellite STPs (Table 4.3) reveal under the current flow regimes and BOD surface loading rates, that Bugolobi, Kyambogo and Unise systems are overloaded, Namboole has reached the full capacity, whereas Naalya and Ntinda systems are underutilised. The design HRTs are within the typical range of 5-30 d for facultative ponds and 5-20 d for maturation ponds (Von Sperling & Chernicharo, 2005), except in Kyambo second pond (4.5) and Namboole second and third pond (4.3 and 3.5). However, the current HRTs show that Bugolobi and Kyambogo second ponds have been reduced drastically to 3.8 and 1.7 d respectively. Namboole ponds receive low flow over long periods from the stadium due to intermittent use of the stadium for national and

Table 4.4. Influent and effluent wastewater concentrations for Kampala satellite ponds (standard deviation between brackets).

Ponds		pH	DO[1] (mg/l)	COD (mg/l)	TSS (mg/l)	NH_4-N (mg/l)	TP (mg/l)	FC MPN/100 ml
Bugolobi	influent	7.1	0.0	819 (149)	377 (123)	39 (9)	17 (10)	3.3E7(1.6E7)
	effluent	7.5	0.6	148 (91)	146 (48)	23 (9)	12 (4)	1.0E4 (3.6E3)
Naalya	influent	7.2	0.0	569 (241)	294 (116)	40 (15)	14 (3)	3.6E7 (3.1E7)
	effluent	8.1	1.0	69 (27)	52 (31)	17 (10)	8 (3)	8E3 (4.0E3)
Namboole	influent	7.4	0.0	745 (96)	621 (183)	25 (13)	10 (3)	1.8E7 (1.6E7)
	effluent	8.1	0.6	130 (70)	143 (61)	7 (6)	5 (4)	5.6E3 (1.8E3)
Ntinda	influent	7.2	0.0	828 (269)	621 (164)	28 (6)	15 (3)	1.8E7 (8.4E6)
	effluent	7.3	0.8	93 (28)	115 (58)	6 (3)	7 (3)	4.4E4 (4.6E4)
Unise	influent	7.1	0.0	581 (192)	371 (104)	27 (5)	13 (2)	2.5E7 (1.7E7)
	effluent	7.3	1.8	56 (25)	20 (11)	5 (7)	5 (3)	3.0E3 (9.3E2)

[1] DO: dissolved oxygen.

international matches. The designed depths of last facultative maturation ponds are on average over a metre deep. Desludging, reuse and recovery practices are lacking.

The monitoring surveys for water quality parameters are depicted in Table 4.4, with overall performance compliance depicted in Figure 4.4 and percentage reduction in Figure 4.5. Namboole,

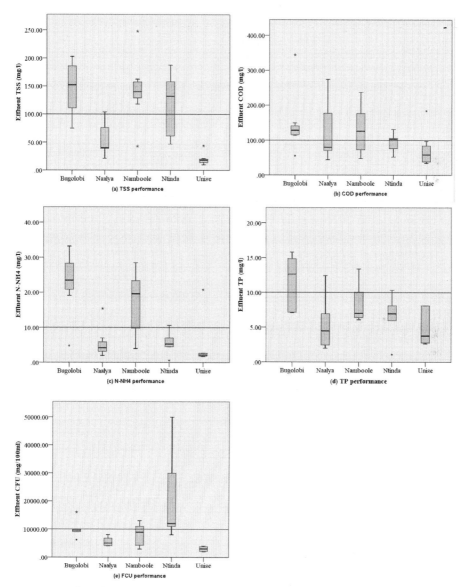

Figure 4.4. Box plot of overall treatment performance regarding (a) TSS, (b) COD, (c) NH_4-N, (d) total P, (e) faecal CFU in Kampala satellite treatment ponds. Concentrations of the various water quality parameters are based on grab samples.
Symbol: Horizontal line indicates discharge standards for respective parameters

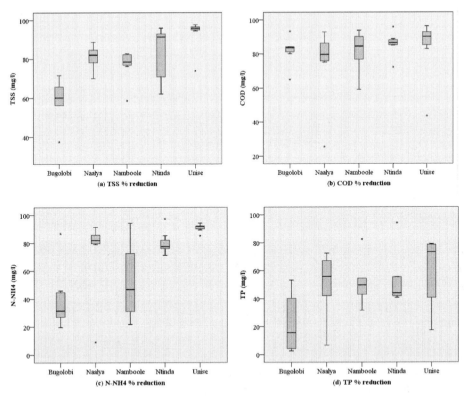

Figure 4.5. Box plots of overall percentage reduction of (a) TSS, (b) COD, (c) NH_4-N, and (d) total P, in Kampala satellite treatment ponds. Efficiencies were calculated based on water quality parameters' grab samples.

Bugolobi and Ntinda have about typical influent COD concentrations, whereas Naalya and Unise have lower. However, since historical data or comprehensive water quality monitoring framework is lacking for satellite STPs, typical BOD generation of 40 g/ca*d adopted for the region is used for the assessments of STPs (NWSC, 2004, 2008; LVSWSB, 2005, 2008; MWI, 2008b). Namboole, Bugolobi and Ntinda have low dissolved oxygen in effluent whereas Naalya and Ntinda have a relatively high effluent pH.

The performance of satellite treatment systems are mixed, some meet effluent discharge standards while others do not (Figure 4.4). Overloaded ponds, e.g. Bugolobi show a weak performance compared to those operated at design capacity, e.g. Namboole and Unise or under design load, e.g. Naalya and Ntinda. From time to time a strong smell is detected from Bugolobi and Kyambogo ponds.

Generally, the low performance of ponds may be attributed to:

- Rapid overloading brought by high population increase or solids accumulations, whereas most ponds have not been desludged since their construction.
- Short-circuiting of flows due to floating objects, vegetation growth and settled solids.

- Ponds coverage by embankment vegetation, duckweed and water lettuce, which prevent light penetration to the ponds.
- Excessive algal growth in the final ponds, which may contribute to high concentrations of TSS in effluents, which together with nutrients (N and P) are then hardly removed in pond systems (Cosser, 1982; Mara, 1996). This is particularly the case when the maturation pond is absent or when it is overloaded, making algae not to settle.
- Use of relatively deep maturations ponds or lack of them in some systems. Most optimal performance is attained in shallow depths of less than a metre down to 30 centimetres (Silva, 1995) compared to over a metre deep in Kampala. For practical purposes, a depth of about 1 m is generally applied.

4.6 Management of existing satellite systems

Satellite sewerage design and construction are undertaken by private firms. In the 1960s and 1970s they were supervised and approved by the city council. From the 1990s onward, DWD has taken over these tasks. There were no records about the sanitary infrastructure, e.g. sewerage plans, files or maps with satellite utility owners, DWD or KCCA, except for the on-going projects like Namungoona and Lubowa 80. Namboole is the only satellite system with a physical office for operation.

Satellite systems are developed by multiple providers, with most developers doubling as operators (Table 4.8; Figure 4.6). Ntinda is managed by Ntinda Neighbourhood Association, Namboole by Sports Management Council (SMC) and Naalya by NHCC. The association operates the satellite utility without management being delegated or ownership handed over to them. Rather, they responded to the deterioration of treatment ponds and sewer blockages due to lack of maintenance. The association manages through a committee meant for welfare and not utility management. The workers are on temporary contract with a monthly wage and no other benefits. Accountability on who to report to and supervision is lacking. Namboole is developed by the state with management delegated to SMC, a body within Ministry of Sports, which reports to the Minister in charge of sports. The directors are appointed by the state. The personnel are paid a monthly salary from government allocation.

Sanitary infrastructure management is not part of developers' core mandate, but as part of estate maintenance, which is not attached to a service fee. Estate maintenance is not done by fully-fledged service units, but workers are assigned general duties as need arise. Where properties have been sold like in residential satellite areas, developers who still claim ownership have little obligation and accountability in their maintenance since there are no other services they offer. Thus their operation is viewed as a favour, a free service to the user community.

There are three personnel staff in Naalya and two in Namboole and Ntinda. The personnel in Namboole and Naalya are permanent and earn monthly salary whereas those in Ntinda are casual workers on an informal contract with the association. Namboole, in addition to paying salary, also accommodates the workers at the treatment site. In Bugolobi, NHCC hires casual workers to clear sewer lines and overgrown vegetation, but the state of the ponds indicated that such an activity has not been undertaken for years. Personnel operating the systems generally lack any formal training in operation and maintenance of wastewater infrastructures. Maintenance regimes

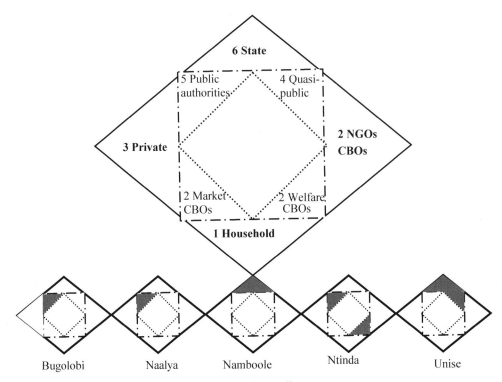

Figure 4.6. Management arrangements in Kampala satellite systems.

such as desludging schedules, expansion plans and service charges and fees are lacking. Treatment ponds have never been desludged since construction with an exception of Ntinda where the first pond was desludged in 2004. Payment of annual discharge fees, monitoring fees and user fees as provided for in the regulation (Uganda, 1995a, 1995b, 1999) is lacking. The DWD did not yet appoint sewerage authorities to manage satellite areas as required by legislation. Consequently, satellite sewerage areas are not operated based on performance contracts, service levels and targets.

4.7 Assessment of satellite system dimensions

Empirical results of satellite systems from Kampala are mapped in this section along the four MM dimensions (Figure 2.3) and assessment scales (Chapter 2, Section 2.5.4; Figure 2.7) to inform on the kind of configuration they exhibit.

4.7.1 Spatial-technical scale: large versus small scale systems

The scales of satellite systems based on population served are community, neighbourhood and small-urban (Table 4.5). The area they occupy ranges from 17 to 331 ha, with planned population size of 1,500 to 14,000 (Table 4.2). Therefore, the assessment scales are 2, 3 and 4 to reflect community, neighbourhood and small-urban service level respectively (Table 4.5; Figure 4.7).

Table 4.5. Assessment scales for satellite systems dimension: large versus small.

Spatial/service level	Population (P) served	Satellite systems	Assessment scale
Community	50-1,500	Unise, Lubowa, Kiwatule	2
Neighbourhood	1,500-5,000	Namboole,Bugolobi, Ntinda, Naalya, Mukono	3
Small-urban	5,000-50,000	Kyambogo, Namungoona, Royal Palms	4

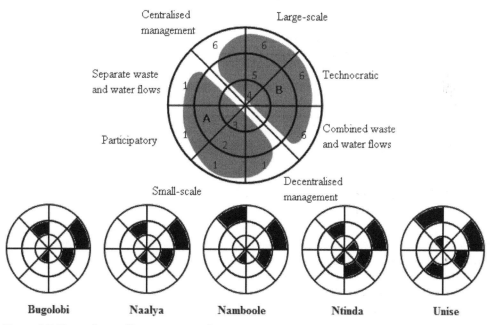

Figure 4.7. Kampala satellite systems configurations.

Satellite areas have a fixed planned population, with infrastructures sized to the planned population. They are decentralised intermediate sewers and treatment systems outside public sewerage areas. However, by 2030, Bugolobi and Ntinda satellite systems will be part of Nakivubo and Kinawataka sewerage catchments respectively (Figure 3.4). This would offer two possibilities for the satellite systems. On the one hand, existing treatment systems can be abandoned and the flows connected to public sewer trunk lines or treatment works in tandem with existing regulatory requirements. On the other hand, satellite systems can be operated as independent sewerage and treatment systems within a public sewerage area.

4.7.2 Management arrangements: centralised versus decentralised

Examining the roles and responsibilities allocated by the regulatory framework (Table 4.1) and situations examined in Kampala, a mix of centralised and decentralised characteristics is discernible in satellite systems (Table 4.6). From Tables 4.1 and 4.6, it follows that some responsibilities are performed by centralised state agents, others are decentralised and some are shared. Management arrangements from Tables 4.6 and 4.7 are graphically presented using the standardised tetragon-shape format depicted in Figure 4.6. The variability/versatility in resulting shaded patterns demonstrates the existing mixture regarding management arrangements (Figure 4.6) or institutional arrangements (Tables 4.6 and 4.7). The satellite provision, production, use and ownership vary, but in general, the form and use of satellite systems can be characterised as private.

Three management arrangements are discernible from Table 4.7 and Figure 4.6, which are developer-operator practised in Bugolobi, Naalya, Mukono and Kyambogo; privately-developed community-managed in Ntinda; and central government-developed user-operated that exist in Unise and Namboole. Except in Namboole and Ntinda, asset ownership and operation and maintenance are not separated. The management scales are 3, 4, 5 & 6 (Table 4.7).

The MM assessment elucidated a flexible mix in management arrangements. The existing mixture in satellite sewerage is due to the diversity of developers and operators unlike urban systems that are provided by monopolistic water and sewerage authorities. A number of inadequacies exist in most of satellite systems due to the management structure of the apparent centralised-decentralised approach in satellite provision (Tables 4.6 and 4.7), which can possibly be ascribed to:

- Lack of appointment of satellite sewerage authorities; sewerage systems are not operated on performance contracts, the instrument that set service targets and standards.

Table 4.6. Assessment of centralised and decentralised aspects in satellite systems.

Aspects	Centralised	Scale	Decentralised	Scale
Technical	• standards, codes and construction protocols	6	• technology choice	4,5,6
Regulation	• supervision	5	• regulation by by-laws	Absent
	• approvals	5,6	• -	-
	• permitting and licensing	5,6	• discharge permit	Absent
	• monitoring	6	• effluent analysis	Absent
	• tariff regulation	6	• tariff indexation	Absent
Operation & maintenance	• appointment of sewerage authorities	6	• utility operation and maintenance	2,3,4,5,6
	• monitoring of performance contracts	6	• performance contracts	Absent
Development	• one-off state support	6	• private, public companies, quasi-public and state	3,4,5,6

Table 4.7. Assessment of Kampala satellite provision, management and user regime.

System	Land use	Developer	Operator	Form of service provision & use	Infrastructure ownership	Scale
Bugolobi	Residentia	NHCC	NHCC	Private	Public	5
Kyambogo	University	University	University	Private	Quasi-public	4
Mukono	University	University	University	Private	Private	3
Unise	Institute	State	Institute	Private	Quasi-public	4&6
Naalya	Residential	NHCC	NHCC	Private	Public	5
Namboole	Stadium	State	SMC	Private	State	6
Ntinda	Residential	NHCC	Association	Private	Public	5

- Sanitary infrastructure management is not part of the core mandate of satellite developers, and therefore, they cannot be accounted for not having the required technical and managerial capacity.
- The community of users being served by satellite systems lack ownership, responsibility and accountability since there are no contractual arrangements in place. Such agreements are necessary since they define roles and responsibilities of providers and users as well as code of practice. Besides, they still view satellite utilities as belonging to the developer.
- Absence of service charges, thus lack of operation and maintenance budgets.

4.7.3 End-user participation: participatory versus technocratic

In satellite systems end-users do not participate in planning, design and construction, which are undertaken by private companies. Besides they are neither informed nor consulted. End-users participation is dismal, except in the operation and maintenance of Ntinda which is in the hands of a neighbourhood association. A framework for end-user participation of households, communities or NGOs is absent. The residents occupy the settlements once developments are completed without participating or making choices, except on the type of properties they want to buy. Satellite systems in Kampala are assessed at scale 4 in operation and maintenance in Ntinda and the rest at scale 6. Therefore, satellite systems can be characterised as technocratic. From the foregoing arguments, satellite systems:
- Have very low end user participation. There are no roles for households or civil society groups. Though they are decentralised systems, they do not necessarily encompass higher community participation opportunities.
- Apply standards, technologies and construction protocols developed for large-scale systems.
- Choice is at the level of purchase for residential properties or affiliation for institutions and government facilities, which pulls people of the same category into a community of users.
- Do not benefit from government subsidies or cross-subsidisation arrangements.
- Are gated through enclave planning, development and user community. It is closed, with those inside enjoying services while those outside are locked out.

4.7.4 Sanitary flows: separate versus combined water and waste flows

Wastewater flows in all satellite areas are domestic sewage and thus configurations are similar, assessed at scale 5 (Figure 4.7). Storm water and sewage flows are separated in design as well as in practice. Storm water is drained by open drains and sewage by closed drains. Cross connection of storm water with sewage flows via inspection chambers, faulty manholes and infiltration cannot be ruled out entirely, but is assumed not significant to warrant mapping. The sewage flows are separated from generation sources based on land use, e.g. residential and institutional campuses or they are facility specific in the case of stadiums and prisons. Further separation of sewage into its constituent flows is absent in all cases; so do reuse and resource recovery practices; bio-solids recovery and reuse are absent. Satellite treatment systems do not receive septage from onsite systems.

At intermediate scales such as in satellite areas, recovery and reuse of resources from domestic sanitary flows is in principle possible given the small area they cover, the leafy suburbs they occupy and their close proximity to agricultural farms. However, currently recovery and reuse is dismal. Satellite systems, nevertheless, have demonstrated that it is possible to collect separate flows from a specific land use or facility generating them and to treat the flows in small decentralised systems. Limited separation of flows can mean limited complexities and less attention from end-users. The nature of satellite development is such that investors want to recoup their investment as soon as possible. Thus the existing flow regime seems appropriate since introducing more separation will require more attention by the end-user community. However, it requires time to sensitise and educate the end-user community on satellite system management.

The choice of waste stabilisation ponds as treatment technology coupled to the satellite sewers is apparently driven by low capital costs, low operational costs, little maintenance, e.g. desludging at 10 to 20 year intervals and availability of (cheap) flat land (Crites & Tchobanoglous, 1998; Sasse, 1998). However, pond systems occupy large areas of land in the city and are not flexible in terms of population growth. Overloading rapidly occurs when population density increases. Moreover, ignoring ponds maintenance in terms of desludging and required repairs will result in malfunctioning systems. The absence of anaerobic ponds in most treatment systems seems advantageous in the direct vicinity of the populated areas, since malodour nuisance is avoided. On the other hand, absence of anaerobic ponds demands for large sized facultative ponds for meeting effluent criteria. Furthermore, absence of maturation ponds in decentralised satellite systems constraints the required pathogen removal capacity as demonstrated in Figures 4.4 and 4.5.

4.8 Conclusion

Satellite systems are developed by multiple providers in peri-urban and rural suburbs independent of public sewerage, are tailor made and financed locally. In term of scales, they are community-sized to small-scale urban. The management arrangements in satellite systems are mixed: Namboole is managed by the state, Ntinda by a mix of public enterprise and a CBO, Bugolobi and Naalya by a public limited company, whereas Unise is quasi-publicly managed with state support. Satellite systems are considered technocratic as they are conventionally designed and constructed by experts without households and civil society involvement. The robustness of sewers in satellite areas, despite the apparent poor maintenance regime, may be attributed to the application of conventional standards and construction codes. Satellite systems have demonstrated that it is possible to separate urban drainage flows from sanitary flows, whereby the sanitary flows are treated in small systems and still achieving effluent discharge standards in the majority of cases. The surveyed satellite systems are based on conventional water-born sanitation principles with no or very limited possibilities for recovery and reuse of resources from the sanitary flows. The recovery and reuse paradigm is apparently not required to guarantee to current functionality of the system. In the surveyed systems, the limited separation of flows resulted in limited complexities, likely contributing to system robustness. The latter is of high importance since satellite systems are characterised by a dismal maintenance regime, less attention from users. They suit the middle and high income classes that aspire for the convenience of water-based flush toilets systems. The applied decentralised treatment systems are conventional and mainly based on pond technology. Current experiences indicate that satellite systems, if adequately managed, can be a viable parallel intermediate sewerage systems provision in East African cities that complement urban public sewerage systems.

Chapter 5.
Onsite sanitary provision as transient or permanent solution in East African cities

5.1 Introduction

Meeting the MDG sanitation target in East African cities is a big challenge and acceleration of onsite sanitation provision is inevitable for any chance to come close by. Despite dominance of onsite sanitation in East African cities (Table 1.2), it is often seen by local authorities (LAs) and sewerage agencies (SAs) as a transient solution to be replaced by sewerage. Onsite sanitation facilities, moreover, hardly focus on the entire management chain, resulting in poor services delivery while threatening public and environmental health. This has to change if on-site sanitation has to become the foreseen key-player in meeting the MDG sanitation targets in a sustainable way. This chapter posits that onsite sanitation provisions can be a transient or a permanent solution depending on mass flow, spatial requirements, appropriate embedding in entire sanitation management chain, and socio-economic feasibility. However, for better sanitation provision, a permanent solution, with room for amendments to anticipate changes in space, mass flow, and sanitation chain management is imperative. To support this statement, this chapter is built up as follows: after briefly explaining the data gathering approach and methodology for this chapter (in Section 5.2), all available onsite technology options in Kampala and Kisumu are discussed in Section 5.3. In Section 5.4 the faecal sludge management practices from emptying latrines to treatment and reuse are presented. Then the institutional arrangements found around onsite sanitation options are discussed in Section 5.5, followed by a discussion on onsite sanitation as a transient or permanent solution. Lastly, the assessment for the onsite systems along the four dimension of the MM is done (5.7) and conclusions presented (5.8).

5.2 Approach and methodology

Assessment of onsite sanitation is done in two ways. First, assessing whether onsite sanitation solutions should be classified as transient or permanent based on population density (P/ha), base flow density (m^3/ha*d), spatial requirements and faecal sludge emptying practices. Second, assessment of onsite sanitation configurations utilising MM dimensions (Figure 2.3) along four axes and 6 level assessment scales as defined in Chapter 2, Section 2.5.4. Mapping of sanitary dimensions is done by way of shading between four axes and three concentric lines whereas assessment is done on a scale of 1 to 6. To achieve the assessments above:
- Data were derived from primary data collection through field visits during which observations and interviews took place. Interviewed stakeholders comprised of onsite sanitation project officers, public health officers, emptying service providers, sewerage personnel, voluntary sector organisations officials, and community and public pay toilet operators. Secondary data were obtained through archive retrieval from onsite sanitation service providers' records, sewerage technical reports, sectoral reports; and acquisition of laws, regulations, and guidelines.

- Qualitative data are analysed through content analysis whereas quantitative data are analysed through excel and presented in tables and charts.
- Secondary data are triangulated with primary data.

5.3 Technology options

5.3.1 Onsite sanitation types, distribution and trends

The dominant onsite sanitation in Kampala and Kisumu are traditional pit (TP) latrines followed by septic tanks and ventilated improved pit (VIP) latrines (Figure 5.1). Other options available but in limited use are ecological sanitation (eco-san), biogas latrines (bio-latrines) and bucket latrines. Shared sanitation account for about 37% and 31% in Kampala and Kisumu respectively (Figure 5.2; Table 5.1). Trends in Kampala indicate that over the next two decades, septic tanks

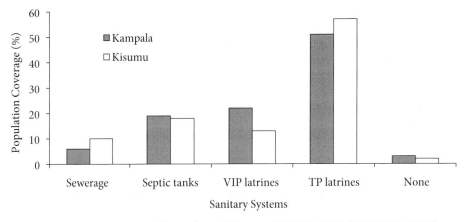

Figure 5.1. Sanitation coverage in Kampala and Kisumu (NWSC, 2008; KIWASCO, 2008).

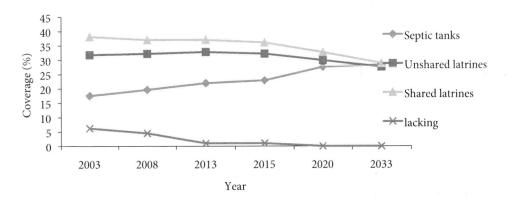

Figure 5.2. Kampala onsite sanitation coverage trends (NWSC, 2008).

Table 5.1. Percentage distribution of onsite sanitation in Kisumu (KIWASCO, 2008).

Urban structure		Onsite household system				Public toilets[1]		
Sub-location	Category	Septic tank	VIP latrine	TP latrine	Shared latrine	Market	School	Health centres
Township	urban	70	15	10	5	169 (1)	210 (23)	84 (7)
Township-Kaloleni	urban	0.6	0.5	2.4	96.5	8 (4)	51 (32)	12 (3)
Kibuye-Migosi	urban	52	26	12	10	8 (2)	16 (4)	21 (7)
Kibuye-Nyawita	urban	16	20	26	38	6 (2)	123 (6)	4 (3)
Milimani	urban	80	15	5	0	42 (1)	79 (8)	37 (5)
Kanyakwar	urban	5	10	25	60	0 (1)	10 (3)	2 (2)
Nyalenda	peri-urban	10	15	25	50	14 (2)	18 (11)	0 (3)
Manyatta	peri-urban	17	5	40	38	3 (2)	15 (5)	12 (4)
Wathorego	peri-urban	29	20	46	5	0 (6)	3 (7)	3 (4)
Korondo	peri-urban	2	7	61	30	10 (3)	25 (5)	9 (3)
Kogony	peri-urban	3	7	20	70	0 (3)	15 (6)	1 (1)
Kasule	rural	17	42	30	11	2 (3)	37 (8)	0 (4)
Chiga	rural	0	0	25	75	6 (4)	12 (3)	4 (2)
Nyalunya	rural	5	35	40	20	4 (4)	7 (12)	1 (12)
Kodero	rural	0	10	90	0	0 (4)	25 (6)	4 (2)
Got Nyabondo	rural	0	1	80	19	0 (3)	25 (6)	2 (2)
Konya	rural	26	23	47	4	0 (4)	1 (0)	3 (5)

[1] Between brackets the number of markets, schools or health centres accounting for respective public toilets.

coverage will increase steadily, while shared and unshared latrine coverage will be decrease and reach an equilibrium in 2030 (Figure 5.2).

The distribution of onsite sanitation in Kisumu (Table 5.1) indicate that in urban areas, septic tanks are dominant; in peri-urban areas, they are a mixture of septic tanks, VIP latrines, TP latrines, and shared facilities; in rural areas, they are mostly TP and VIP latrines; whereas in slums, they are mainly shared sanitation facilities. Sanitation in public places such as markets, schools and health centres are inadequate.

5.3.2 Septic tanks

Septic tank systems serve (a) individual household housing units, (b) apartments on single standard plots, (c) apartment clusters and (d) groups of households by way of shared sanitation. Septic tanks in individual households generally consist of two chambers, with the second chamber being the soakage pit. In most rental apartment housing, septic tanks have soak pits, but are not performing as required due to inadequate emptying frequency, poor design and insufficient area for percolation. Soak pits in such circumstances are often surcharged and overflowing. In some instances, bathroom and kitchen wastewater are conveyed to open drains along access roads, which together with overflowing sullage from soak pits, result in streams of greyish to black water flowing throughout the year in such areas.

In densely populated slum settlements, septic tanks on communal sanitation blocks are applied with more than two chambers. For instance, the sanitation block on septic tank in Manyatta slum by an NGO (SANA) designed to serve 500 P/d, has three chambers. In Namungoona Phase I apartment cluster with 144 dwelling units and population of about 800, four chambers comprising of two septic tanks and soakage pits in series are used as a transient solution to be abandoned when satellite treatment system is established at a later date.

There are some residential areas using septic tanks within sewerage areas such as Nakasero 3 and Naguro 1 in Kampala (Table 3.4) and Milimani and Migosi (Kibuye) in Kisumu (Tables 3.5 and 5.1). In these areas, septic tanks that operate since colonial time have soakage pits, and average plot sizes of about 900 m^2, with the lowest being 450 m^2. They are well sited and construction is supervised by the council. However, Milimani North, Nakasero 3 and Naguro1 are undergoing land use change from low density residential to commercial and residential apartments, resulting in high densities, and thus have become a target for sewerage connection (Tables 3.4 and 3.5).

5.3.3 Pit latrines

TP latrines are the dominant excreta disposal method in Kampala and Kisumu rather than VIP latrines (Figure 5.1; Table 5.1). Raised pit (RP) latrines dominate Kampala valleys where settlements are located on reclaimed wetlands with high water tables. In Kisumu's high-water table areas like Manyatta, Nyalenda and Kogony (Bandani) where water level rise to about 3 m to ground (LVSWSB, 2005a, 2008), pit latrines are less than 2 m deep and unlined. There are on-going projects for promotion of lined VIP latrines by KCCA under Kampala urban sanitation project (KUSP) and LVSWSB under short term action plan (STAP). The lined VIP latrines are about 3 m deep, lined with concrete blocks reinforced with steel bars, constructed in poor or high water table areas and they lack soakage pits.

Latrine superstructures are constructed from mainly bricks. However, stone, mud and pole, iron sheets and even plastics are also in use.

The study found that in more densely populated peri-urban areas; pit latrines are shared, heavily loaded, poorly built and badly maintained. In particular, household pit latrines in slum settlements are poorly constructed, walls and roofs are in varying stages of collapse, doors are usually lacking and space is limited. They are full or nearly full and in dilapidated state. Once the latrines are full, sanitary disposal becomes impossible, with filled pits simply left to overflow into the environment or manually emptied. In Kampala valleys, pit latrines are raised and constructed with holes at the side. In Kisumu high-water table areas, shallow pit latrines are used. In sparsely populated rural areas, pit latrines are generally in good condition, operated well, serve a household and built to considerable depths where groundwater levels are low and soils are stable. The design and construction problems associated with pit latrines in Kampala and Kisumu are:
- they are unlined, thus susceptible to collapse when emptied;
- poor ventilation and difficulties in cleaning slabs, leading to odour problems;

- large holes, making it difficult for children to use for fear of falling into;
- flooding during rainy season;
- poor quality of pits and superstructure construction;
- inadequate information on appropriate sanitation, design and cost;
- some contractors lack necessary skills and experience to build adequate sanitation facilities;
- RP latrines are noted to be difficult to use by children, elderly, sick and pregnant women.

5.3.4 Ecological sanitation (Eco-san) latrines

Ecological sanitation (Eco-san) is dry dehydration urine diverting system, which recycle nutrients in stabilised urine and excreta. When properly designed and operated, eco-san systems can provide a hygienically safe, non-polluting and cost-effective sanitation solution (Wikipedia, 2012). Eco-san project in Kampala targets poor settlements with limited productive assets and have incomes far below the average city population GDP of €225 per annum, and where sewerage connection is inaccessible. High water tables make pit latrines and septic tanks unsuitable, high demand for shared sanitation facilities and areas with supportive leadership (NWSC, 2008;Carlesen *et al.*, 2008). Eco-san projects target 20 parishes under Kampala ecological sanitation project (KESP). So far the toilets have been implemented in five parishes: Kamwokya, Bwaise, Wabigalo, Kasubi, and Kyanja. The households and communities initial cost are subsidised by about 95%. KESP was expected to develop and test five alternative eco-san toilets, but one prototype was developed costing about €800. The designed prototype and applied subsidies are considered too expensive and unsustainable (Carlesen, Vad, & Otoi, 2008). The time for storage of faecal matter is 3 to 4 months. The link between eco-san toilets and urban farmers is yet to be established. Eco-san designs are perceived technically sound with 2-4 vaults chambers.

Eco-san toilets in Kisumu are located in the slum settlements of Nyalenda and Manyatta. The first eco-san projects were implemented in Manyatta by an NGO (SANA) targeting households. During field survey, only one toilet was still in use, but utilised more as a showcase since the household had a VIP latrine. Five eco-san toilets had been constructed in Nyalenda by September 2011 by a consortium[20] of NGOs, but were yet to be used since community awareness was on-going. The toilets have two chambers and will be used as communal toilets.

5.3.5 Biogas latrines (Bio-latrines)

Bio-latrines is a relatively new sanitation technology and is applied in Kisumu slum settlements. There are five bio-latrines in the slums of Nyallenda (1), Obunga (Kanyakwar) (1), Manyatta (2) and Bandani (Kogony sub-location) (1). Nyalenda bio-latrine serve Pand Pieri Primary School and surrounding community, with a bio-digester dome of 54 m^3, gas dome of 18 m^3 and design population of 600 P/d: school 400 P/d and community 200 P/d. The number of stands for the school is 18 and that of the community is 9. The system is a pour flush with one litre bucket of water used for flushing. The community pays about € 0.02 per visit to the toilets and about € 0.05 per shower. There is no post-treatment of effluent that flows to the environment. Earlier attempts

---

[20] Practical action East Africa, KUAP and Shelter Forum.

to post-treat with constructed wetlands were unsuccessful due to clogging. The school plans to utilise the biogas for cooking and boiling water once the kitchen is relocated to about 60 m from the bio-latrine. At the moment, the school kitchen is 130 m from the bio-latrine. The bio-latrine in Obunga has a bio-digester dome of 31 m^3, gas dome of 11 m^3, and 9 stands toilet facilities, 4 for gents and 5 for ladies. Manyatta bio-latrine has a bio-digester dome of 31 m^3, expansion chamber of 18 m^3, gas dome of 18 m^3, and 9 stands toilet facilities, 4 for gents and 5 for ladies.

Bio-latrines are planned such that they serve residents within about 60 m radius. The rational for the 60 m radius is that at this range:

- the user community members know each other and thus it is easy to mobilise them during construction, operation and maintenance;
- piped biogas distribution for use can be secured and safety ensured;
- accessibility and security through social and spatial proximity is tenable;
- toilet block can be coupled with community basic facilities, amenities and services, e.g. in Kisumu they are planned as community centre with a range of other services such as water kiosk, shower facilities, offices, community hall, restaurants, stalls and community banking hall.

The bio-latrine technology adopted is a fixed dome. Fixed dome digesters are considered by the designers as long lasting, need least maintenance when constructed well, are robust with reliable performance and have a 20 year design life.

5.4.5 Shared sanitation

Shared sanitation use is high (Figure 5.2; Table 5.1), with trends showing they will remain a significant mode of provision over the next two decades. Shared sanitation occurs mainly in high-density slum settlements and low-cost rental housing. The average number of persons sharing a sanitation facility in Kampala is estimated at 35.5 (6.5 households) per stand (NWSC, 2004). In rental housing, households share central toilets whereas in slum settlements, households share communal toilets. Public or community onsite sanitation facilities have multiple stands ranging from 4-8, with exceptional cases like 24 stands toilet block at Kamwokya primary school in Kampala.

5.4 Faecal sludge management practices

5.4.1 Emptying

Faecal sludge emptying practices in Kampala and Kisumu are either mechanical or manual. Mechanical emptying is done by cesspool vacuum trucks. Emptying distance depend on the length of the hose pipe, which vary depending on pump power. The average length per service provider category are: council – 20 m in Kisumu and 20-50 m in Kampala; private – 20-50 m in Kisumu and 50-100 m in Kampala, with two vehicles having about 150 m hose pipes in Kampala, but they could only empty sludge within 100 m; and private institutions have mainly 20 m long hose pipes. Mechanical emptying trucks in Kampala and Kisumu empty septic tanks, followed by VIP latrines, then public toilets (Figure 5.4). TP latrines are rarely emptied mechanically due to risk of collapse and need for water to make it fluid to enable sucking of sludge. Mechanical emptying costs are lumped together with collection costs (Table 5.2).

Manual emptying is undertaken by a cadre of independent manual emptying service providers referred to as 'scavengers' in Kampala and 'scoopers' in Kisumu. During manual emptying, the scavengers in most cases are drunk; they undress completely, smear their bodies with used engine oil and operate at night. It is done in groups of 2 or 3 and wages shared. Manual operators tools comprise of buckets and long construction poles or adjustable metal bars, the latter practised by a few manual emptying operators in Kampala. Manual emptying is mainly in inaccessible settlements and comprise of (a) scooping with the bucket and pouring the contents into drainage channel, (b) digging of a pit beside the latrine if space is available, emptying the sludge contents into the pit and covering with the excavated soil or in some cases left uncovered, and (c) emptying of bails in nearby undesignated sewer manholes in parts of Kampala Central and Makindye Divisions despite officially being phasing out for decades. The cost of adjustable metal bars is about €120 and timber poles (four required) is about €24. Manual emptying service providers are paid a fee ranging from €2 to €8 in Kisumu and €3 to €9 in Kampala, with payment depending on the informal negotiations and willingness to pay.

Another category of manual emptying is where households open holes from the sides of RP latrines during rainy season to discharge excreta. The storm water is used to flush away the excreta.

5.4.2 Collection

The number of cesspool vacuum trucks per type of service provider dedicated for faecal sludge collection services are: five trucks owned by the council in Kampala and two in Kisumu; 38 private trucks in Kampala and two in Kisumu; and five trucks owned by institutions in Kampala. Vacuum truck capacities range between 2 m^3 and 11 m^3, with only medium-sized trucks (5 m^3) operating in Kisumu (Table 5.2). The amount of faecal sludge collected in Kampala, based on dumping records for 2004-2008 at Bugolobi, averages about 200 m^3/d (4,000 m^3/month), whereas in Kisumu it is about 18-32 m^3/d based on field interviews and observations at the last manhole to Nyalenda ponds in August and September 2009.

Table 5.2. Faecal sludge emptying and collection costs in Kampala and Kisumu cities.

City	Vehicle size and standard cost of latrine emptying			Cost per cubic metre of waste (€)
	Small	Medium	Big	
Kampala (€)	13	25[a]	33	3-7
Kisumu (€)	-	30	-	5

[a] During survey Kampala city council charged about €17 instead of €25 standard rates; 1€ = 100 Ksh and 3,000 Ush.

Faecal sludge collection estimate for Kampala are septic tanks (60%), VIP latrines (36%), TP latrine 3% and public toilets (1%) (NWSC, 2008). Faecal sludge collection in Kampala lags behind accumulation rates (Figure 5.4). The gap between the theoretical volume and actual collection in 2008 was estimated at 350 m³/d, with trends expected to remain so over the next two decades (NWSC, 2008).

The costs of faecal sludge emptying and collection in Kampala within 8 km radius depend on the size of the vacuum truck used, whereas in Kisumu it is uniform: only medium sized trucks are used and the service distance is generally within a 5 km radius (Table 5.2). The variable costs in Kampala are based on distance above the 8 km radius, length of hose pipe used and amount of water used to dilute the sludge. Faecal sludge emptying and collection costs per cubic metre are the same for Kampala and Kisumu if medium sized trucks are compared.

Operators of private vacuum trucks are members of Private Emptiers Association (PEA) in Kampala whereas in Kisumu, they are too few to be associated. In Kisumu private vacuum trucks do register with the council through payment of about €30 per year to operate in the city. In Kampala, they registered with PEA through payment of €70 association fee and daily fee based on vehicle size: €0.2, €0.4 and €0.7 for small, medium and large respectively. The PEA members benefit from free parking space, security of vehicles, price regulation, visibility and welfare needs. Vacuum trucks managed by KCCA were donated as part of KUSP and distributed one per division, but during the field survey, only three out of five trucks were operational. Other institutions with vacuum trucks for their exclusive use are the police, the army and Kampala International University with one each and NWSC with two. Kisumu had one vacuum truck disused beyond repair.

The challenges facing faecal sludge emptying and collection in Kampala and Kisumu can be grouped into two, those facing manual scavengers/scoopers and mechanical vacuum tankers. Manual services providers are faced by lack of suitable emptying, collection and disposal infrastructures; besides they are outlawed and thus illegal, insanitary, and pollute the environment. Mechanical service providers are hampered by lack of access roads, long distance to designated centralised tipping points, high cost of transport[21], use of solid materials for anal cleaning or disposal of solids into pits that tend to block sucking of sludge[22] and tipping restriction to normal working hours, 8 am to 5 pm.

5.4.3 Tipping

Faecal sludge is tipped at designated sites, Bugolobi central STW old humus tanks in Kampala and the last manhole to Nyalenda WSP in Kisumu. The tipping fee in Kampala is about €1.8, 2.5 and 3.5 for vehicles of <2 m³, 2-5 m³ and >5 m³ respectively. KIWASCO is charging an annual fee of about €120 per vehicle as a tipping fee.

---

[21] Cesspool trucks have to pass through the city centre, which is bedevilled by traffic jams while criss-crossing the city in search of customers or when tipping at STPs.

[22] A vacuum truck collapsed in 2007 in Kampala when the hose pipe was blocked while exhausting, leading to blockage, which led to build up of pressure inside vacuum tanker causing its collapse.

5.4.4 Characteristics of onsite sludge

Analysis of faecal sludge characteristics in Kampala shows that they vary from one sanitation option to another (Table 5.3). Faecal sludge from septic tanks are more diluted and of lower concentrations than that from latrines and public toilets. Faecal sludge from septic tanks and TP latrines are well stabilised (45% TVS) whereas faecal sludge from RP latrines and VIP latrines are less stabilised. Those from public toilet exhibit characteristics of high strength fresh sludge with high TS and NH_4 content and relatively low COD/BOD ratio.

The high concentration from TP latrines is because they are not diluted with water except during emptying, where urine and any used water has soaked away. The stabilised sludge from septic tanks and TP latrines are attributed to the low frequency of emptying, which is about every 3 years compared to VIP latrines that are emptied on average yearly or RP latrines that are emptied biannually. Lined VIP latrines are less stabilised as they are emptied more frequently due to low storage capacity. The high strength sludge from public toilet is because they are emptied more frequently due to the high usage, estimated at every 1-2 months (NWSC, 2008).

5.4.5 Treatment and reuse

The faecal sludge in Kampala is tipped into an open inlet channel at an old humus tank, which then passes through a screen and flows into the first chamber where solid material settle and liquid flow into a second chamber. From the second chamber, pre-settled effluent flow into lower pumping station where it is transferred via pumping to central Bugolobi STPs inlet, to be subsequently co-treated with municipal sewage. Faecal sludge in Kisumu is co-treated with municipal sewage at Nyalenda ponds without recovery and reuse of bio-solids. Kisat STPs used to receive faecal sludge

Table 5.3. Faecal sludge characteristics in Kampala (NWSC, 2008).

Parameter	Unit	Septic tank	VIP latrine	Pit latrine	Raised pit latrine	Public toilet	Average quality
TS	g/l	22	30	40	30	35	31.4
TVS	g/l	9.9	19.5	18	18	24.5	18
TVS	% TS	45	65	45	60	70	57
COD	g/l	10	30	35	30	30	27
BOD	g/l	1.4	5.5	5	5	6	4.6
COD/BOD	Ratio	7.1	5.5	7	6	5	6.1
TKN	g-N/l	1	3.4	5	3.4	3.75	3.3
NH_4	g-N/l	0.4	2	2.5	2	3	2
TP	g-P/l	0.15	0.45	0.5	0.45	0.4	0.39
Helminth eggs	No. x 10^3/l	4	30	40	30	30	26.8

before decision was made to centralise tipping at Nyalenda ponds in 2003 due to overloading in Kisat STPs and underutilisation of Nyalenda ponds.

There are plans to establish faecal sludge treatment plants (FSTP) in Kampala in Lubigi by 2013 and Nalukolongo and Kinawataka by 2023 to supplement Bugolo/Nakivubo STPs (NWSC, 2008). The FSTP will also treat sewage in the future when sewerage is established (NWSC, 2004, 2008).

5.5 Institutional arrangement

Many stakeholders provide onsite sanitation services in Kampala and Kisumu (Table 5.4). Nearly all private or shared household latrines are provided by households and landlords. Where tenants lack adequate sanitation facilities, private entrepreneurs do built pay toilets. The public agencies involved in sanitation are KCCA, MCK, NWSC and LVSWSB, either funded from taxes or by external development partners. They provide public toilets in public places, e.g. markets, health facilities, schools and community toilets. NGOs and CBOs are involved in the installation of community sanitation blocks and hygiene promotion; whereas Ministries are responsible for regulation and hygiene promotion. Maintenance is generally carried out by the users and by operators in pay toilets.

The forms of sanitation services provision are:
- *Household sanitation services*. Latrine construction, use and management vest entirely with the households concerned. They are also responsible for making arrangement for emptying

Table 5.4. Onsite sanitation provision matrix in Kampala and Kisumu.

Forms of provision	Service providers[1]									
	HH	LL	PE	SAs	LAs	NGOs	CBOs	Scavengers	MoH	MoE
Household latrines	X	X				X				
Public latrines					X	X				
Private toilets			X							
Community latrines					X	X	X			
School latrines				X	X					X
Manual emptying								X		
Mechanical emptying			X		X					
Septage treatment				X						
Regulation					X				X	
Enforcement					X					
Health promotion					X	X	X		X	X

[1] HH: households; LL: landlords; PE: private entrepreneurs; SA: sewerage authorities; LAs: local authorities; NGOs: non-governmental organisations; CBOs: community based organisations; MoH: Ministry of Health; MoE: Ministry of Education. The shaded cells indicate services that are provided by the respective service provider.

once they are filled up. NGOs do promote household sanitation facilities as demonstration projects for adoption by households.

- *Public sanitation services.* These are public toilets in markets, bus parks, schools and health centres provided by LAs or the SAs. LAs offer four forms of public toilets services: (1) public authority owned and operated pay toilets, (2) public authority owned and private entrepreneur operated toilets, (3) pubic authority owned and CBO operated toilets and (4) public authority developed and user owned and operated toilets. In bus parks, users pay about €0.1, whereas in user owned and operated facilities, toilet use is free of charge. In publicly developed school and health centre sanitation facilities, ownership and management are handed over to the respective administration.
- *Private sanitation services.* Sanitary blocks in informal slum settlements are based on paid toilet services. Private entrepreneurs develop, operate and manage the toilets, with users paying about €0.02 per visit for toilet use and about €0.05 for shower.
- *Community sanitation services.* These are community sanitation blocks in informal slum settlements. They are developed through NGO-community partnership where the NGO raises money from external development partners and the community contribute by way of labour and land. The installed facilities are managed by marketized CBOs as pay toilets. Besides latrines, community sanitation blocks include a small shop or water vending kiosks to secure financial sustainability.
- *School sanitation services.* Ideally schools construct their own facilities, but where they are unable to do so, LAs, SAs and even central government may assist. The responsibility for the management of sanitation facilities installed in schools, in all cases, lies with the school management.
- *Latrine emptying services.* In general, LAs have the overall responsibility for ensuring the provision of efficient exhauster services as the health authority for the city. However, private entrepreneurs are dominant in mechanical emptying, which is the formal market; whereas manual emptying services are dominant in the informal market. Some institutions do have cesspool vacuum trucks for their own use.
- *Regulation and enforcement services.* The roles and responsibilities assigned to key institutions by legislation (Uganda, 2000, 1995a, 1997a; Kenya, 1986, 2002) and environmental health policy (MoH, 2005; MoH, 2007), are that Ministries formulate policies, laws and regulations whereas LAs and public health officers enforce them. The SAs are responsible for septage treatment. Local Governments Act (Uganda, 1997b; Kenya, 1998) places prime responsibility on LAs to develop by-laws for provision of onsite sanitation, faecal sludge and development control[23] services; and to ensure their compliance.
- *Health promotion.* The sanitation improvement efforts of public health authorities only advise user groups on the appropriate technologies and designs for their situation. This advice is only given when sought and upon payment, thus not frequently given. NGOs and CBOs promote particular sanitation technology options, offer technical advice, undertake training in latrine constructions, and capacity building. Government agencies offer health promotion through

---

[23] For instance, every development within jurisdiction of a local authority is supposed to apply for a building permit and submit plans, architectural, structural (in case of high-rise) and sanitation plans for approval.

mainly water and sanitation hygiene (WASH) approach whereas the approaches of NGOs and CBOs are differentiated, i.e. child hygiene and sanitation technology (CHAST), child to child (C2C) and personal hygiene and sanitation transformation (PHAST), in Kisumu.

5.6 Onsite sanitation as permanent or transient solution

The general sanitary engineering approach espoused in the Kampala sanitation master plan and draft sewerage manual for Kenya (NWSC, 2004; MWI, 2008b) can be interpreted as one dimensional determined by population and base flow density. Things that are not adequately covered in this approach are:
- Embedding of sanitation solutions in local socio-economic circumstances;
- Embedding of sanitation in the entire water chain, including conveyance, treatment, discharge and possible valorisation of waste constituents;
- Embedding of sanitation in the city spatial structure.

It is therefore hypothesised that when considering the complete set of criteria, population density, base flow density, spatial requirements, and excreta management, a specific sanitation solution at a specific location could be considered permanent whereas the traditional criteria based on base flow and population density would classify it as transient. Following this hypothesis, we are introducing other criteria in addition to the population and base flow density to determine the range of possibilities in judging permanency or transiency of onsite sanitation provision.

5.6.1 Population and base flow density

Figure 5.3 depicts the population density trends in Kampala for the period 2008-2033. For the year 2013, three types of sanitation service areas can be distinguished: sewerage (200-500 P/ha), mixed (transition) (100-200 P/ha), and onsite (<100 P/ha); whereas in 2033, sewerage and mixed service areas prevail (Figure 5.3). The base flow density in Kampala between 2008 and 2033 show that roughly three service areas can be defined: sewerage ($\geq$10 m^3/ha*d), mixed (5-10m^3/ha*d) and onsite (<5 m^3/ha*d) (Figure 5.4). Mixed and onsite service areas by 2030 are likely to be served by on onsite sanitation, except the defined public sewerage areas (Figure 3.4) and satellite sewerage areas (Figure 3.4; Table 4.3). Based on population and base flow densities as decision criteria, in many parishes in onsite and mixed service areas in Kampala, the densities will still be low enough for onsite sanitation to function effectively in 2033 (Figure 3.4).

Sewerage threshold for Kisumu is defined at <120 P/ha (MWI, 2008b). The sub-locations in Kisumu that can be sustained by onsite sanitation (<120 P/ha) by 2030 are 10 out of 14 (Table 3.5). The areas that meet the population and base flow density for sewerage are Kibuye, Milimani, Nyalenda and Manyatta sub-locations. In practice these areas are not sewered despite meeting the population and base flow density. A number of reasons can be ascribed to it: (1) informal and unplanned spatial structures, which are unsuitable for conventional sewerage (2) inhabited mostly by the urban poor, which are considered unable to pay for sanitary services from willingness and ability to pay assessments surveys, and (3) lack of funds and political will for improvement of such areas.

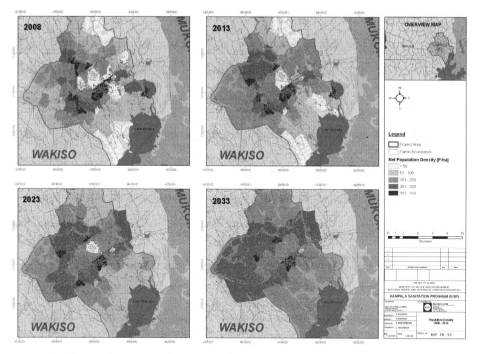

Figure 5.3. Population density trends in Kampala (NWSC, 2008).

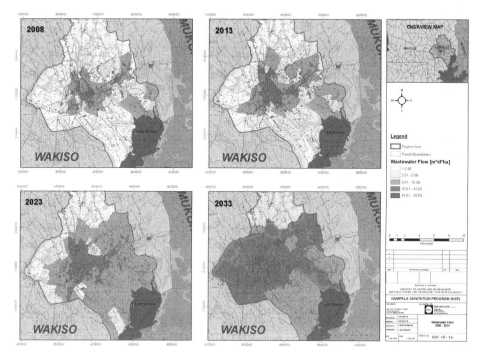

Figure 5.4. Base flow density trends in Kampala (NWSC, 2008).

5.6.2 Spatial and loading requirements

Sanitary regulations, standards, master plans and guidelines provide for spatial, loading, and development requirements (Table 5.5). The onsite sanitation systems that are often approved by LAs are septic tanks since the design and approval criteria is available (Uganda, 2000, MCK, 2008a; Kenya, 1986). In bio-latrines, the structures are approved, but the bio-digester and auxiliary toilet facilities are not due to lack of approval criteria. Consequently, septic tanks and bio-latrines are formal, thus potentially can be designated in the city land use plan and thus can be operated as a permanent sanitary solution in urban areas.

Kampala spatial structure (Figure 3.7) shows that some settlements are located in the valleys, which are reclaimed swamplands with high water table. In Kisumu, many settlements are located in high water table areas, e.g. Kanyakwar, Nyalenda and Manyatta. A high water table hampers or even rules out application of septic tanks, VIP and TP latrines in such areas (MWI, 2008b;MoH, 2000, 2002). From the empirical findings, in these areas, RP latrines are applied and eco-san promoted. In Eco-san, excreta storage is grossly inadequate for safe and hygienic reuse because they are stored for 3-4 months due to high usage against the regulation requirement that faecal sludge be stored for 2 year for safe reuse (MoH, 2000). Eco-san handling in such cases is insanitary and its use unsafe thus can be considered a transient solution which has to be replaced. Yet, when the entire chain is included, e.g. excreta is well stabilised and linkages to reuse of stabilised excreta and urine is established, eco-san latrines can potentially be considered a permanent solution as well.

Septic tanks have been in use in township neighbourhoods established at the turn of 20th century, e.g. Nakasero and Naguro in Kampala and Milimani and Kibuye in Kisumu (Tables 3.4, 3.5, 5.1). In these areas, septic tanks have soakage pits, plot sizes averages 900 m^2 except Kibuye that averages 450 m^2, located in high grounds and sites well drained. An attempt to compulsorily connect these areas, such as the 2004-2006 campaign in Kampala was resisted by households as it was deemed not necessary. This indicate that they still offer good services even at the turn of 21st century, and thus have been operated as a permanent solution. Rezoning of these areas into commercial and apartment buildings, however, have led to densification and concomitant increase in wastewater flow resulting in sewerage thresholds being attained, thus shifting septic tank application from being a permanent to transient solution. This is because of the one-dimensional approach to sewerage provision that applies conventional sewers only. Yet septic tanks could be part of an alternative approach to sewerage being reutilised via small-bore sewer systems conveying the liquid to off-site treatment.

Based on plot arrangements (Figure 3.9a,b) and standard plots of 15×30 m (MoL, 2008;MCK, 2010), while considering the spatial requirements in Table 5.5, septic tanks and bio-latrines can be considered a permanent solution in urban and peri-urban areas and VIP, TP and eco-san latrines in rural areas. In bio-latrines, the bio-digester is constructed below the structure, which deviates from the normal sanitation requirement that latrines be constructed under a building (Kenya, 1986; Uganda, 2000). If the standard plots are on single family housing, the average densities can be 55-110 P/ha. These make onsite sanitation provision, e.g. septic tanks, meet spatial requirement and loading and thus can be appraised as a permanent solution. However, with apartment housing, it results in high densities and thus can be considered a transient solution or otherwise can be part of an alternative approach. So far, in bio-latrines both liquid effluents and solid digestates are not

Table 5.5. Spatial and loading requirements for onsite sanitation in Kampala and Kisumu.

Type/form	Stipulations	Reference
All latrines	• Where a building is not within 61 m of public sewer line	Uganda, 2000
	• Latrines should not be under any building	Kenya, 1986
	• Population density is less than 120 P/ha	MWI, 2008b
	• Population density is less than 200 P/ha	NWSC, 2004
	• Base flow density is less than 10 m³/ha*d	NWSC, 2004
	• Saturation density (P/ha) low 50, medium 250 & high 450	NWSC, 2008
	• A permit from council is required for construction of any latrine for reception or disposal of sewage	MCK, 2008a
Public/shared sanitation	• Maximum loading of 30 P/stand	(MoH, 2002)
	• 25 P/latrine	(MoH, 1987)
	• Convenience and accessible from a street	MCK, 2008d
	• 4 households/sanitation facility	NWSC, 2004
Pit latrines	• 30 m from a well and 10 m from a dwelling unit	(MoH, 1987)
	• 15 m from downstream water abstraction point	MWI, 2008b
	• 30 m from any dwelling	(MoH, 2000)
	• 30 m from existing sewerage lines	(MCK, 2008b)
	• Areas designated by council local physical development plan	(MCK, 2008b)
	• Lining of pit latrines in unstable soils	(MCK, 2008b)
	• Not where water table is within 1 m of ground surface	(MCK, 2008b)
Septic tank latrines	• Septic tanks and drain fields should not be located 30 m from wells and embankments or 3 m from building lines, water points, footpaths and trees	(MCK, 2008b)
Eco-san latrines	• Located in peri-urban and rural areas	MoH, 2000
	• Where groundwater is high and soils are shallow or loose	MoH, 2002
	• Faeces be stored for a period of two years or more to ensure that *Ascaris* die off from the solids	MoH, 2000
Bio-latrines	• Serve households with a 60 m radius catchment	Interview
Exhauster services	• Exhauster services provision requires a permit from council	MCK, 2008a
	• Payment to council or its agent fees and charges per month for regular services of emptying pit or tank	(MCK, 2008c)

adequately dealt with. One condition for permanency is that both liquid effluents and digested slurries are further treated and/or reused in a hygienic and environmentally sound way. If this chain is not adequately covered then these initiatives are doomed to be transient.

Based on loading rates, shared (public, private and community) sanitation blocks accommodate very high densities, e.g.:

- CIDI (NGO) in Makindye Division in Kampala are providing shared sanitation blocks where 50 households share a VIP latrine;
- SANA (NGO) community sanitation block on septic tank in Kisumu's Manyatta slum has capacity to service 500 P/d;
- Umande Trust (NGO) bio-latrines in Kisumu slums serve 200-600 P/d, which if we translate this to a 60 m radius, result to a service provision for 177-530 P/ha;
- apartment buildings on standard plots (15×30 m), and often on septic tanks constructed below structure and serving about 125 P result in 2,778 P/ha;
- Kaloleni, which relies on community sanitation blocks (96%) (Table 5.1) has a density of 1,200 P/ha.

In shared sanitation, sitting is not restricted to a plot. Therefore, it can be located in such a way that they are accessible from a street, loading rates per stand or facility are not exceeded and required distance from water sources, dwellings, sewer lines and embankments can be met (MoH, 1987, 2000; MCK, 2008a, 2008b; MWI, 2008b). Besides, if they are accessible by a street it means they are also accessible by exhauster services. If a shared sanitation fulfils all these requirements, then it can be considered a potential permanent solution. Therefore, at high density, community sanitation can potentially be a permanent solution whereas household solutions can be transient.

Examining onsite sanitation distribution and practices in Kisumu (Table 5.1) reveals that sub-locations with shared (community) sanitation are those settlements (a) located in high water table, e.g. Kanyakwar, Nyalenda and Manyatta; (b) have high population density, e.g. Nyalenda, Manyatta, Kaloleni, Nyawita[24]; and (c) houses slum settlements, e.g. Kogony (Bandani slum), Nyalenda, Manyatta and Kanyakwar (Obunga slum).

5.6.3 Excreta flow

Without emptying, latrines can be considered as improper and therefore transient solutions because they will fill up and are abandoned. Latrine emptying does not necessarily translate into permanency, but has to be coupled with flow of excreta to treatment plants or reuse and recovery practices. In Kampala, septic tanks and VIP latrines account for about 96% of latrines being emptied (NWSC, 2008); with trends showing this will not change much (Figure 5.5). In Kisumu, shared latrines are dominant in high density informal slum settlements (Table 5.1). Besides, in both cities, shared latrines are run as enterprises, with convenience, cleanliness and differentiated service charges to maintain customers and regularly emptied. From the findings from Kampala and Kisumu, therefore, one can postulate that septic tanks and shared sanitation, i.e. bio-latrines

---

[24] Comprise mostly of apartment housing with a density of 647 P/ha based on KIWASCO (2008) population estimates.

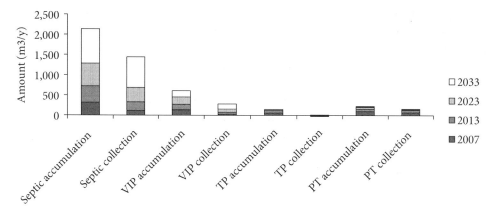

Figure 5.5. Faecal sludge accumulation and collection trends in Kampala (NWSC, 2008). VIP: ventilated improved pit; TP Traditional Pit latrine; PT Public toilet.

and lined VIP latrines, can be considered as permanent solution as they are exhaustible, thus conduit for excreta flow to STPs whereas pit latrines can be transient. In the same reasoning, it is postulated that eco-san, which is currently without adequate storage of excreta or linkage to potential users of excreta, can be considered transient, unless adequate storage and hygienic reuse chain is established.

5.7 Assessment of onsite sanitation dimensions

This section assesses onsite sanitation configurations along the four MM dimensions and 6 level assessment scales and maps their configurations.

5.7.1 Spatial-technical scale: large versus small scale systems

Onsite sanitation solutions are applied at household, apartment, cluster (household or apartment) and community service level (Table 5.6). They are also site, user or catchment specific. The population size ranges from one person in a household to community scale with a population of 500-800. This translates into assessment scale of 1 and 2, with 1 attributed to households and cluster and 2 to community level. Therefore onsite systems are small-scale in nature. However, bio-latrines can accommodate very high densities, up to 500 P/ha against transition density for sewerage of 200 P/ha.

5.7.2 Management arrangements: centralised versus decentralised

Onsite sanitation management arrangements, as expected from the variety of providers (Table 5.4), are mixed (Figure 5.6). Households (assessed at scale 1) and private entrepreneurs (assessed at scale 3) provide pit latrines and septic tanks, with private entrepreneurs provisioning in rental housing and shared private pay toilets. Public authorities (assessed at scale 5): KCCA, MCK and

Table 5.6. Spatial-technical scales of onsite sanitary systems in Kampala and Kisumu.

Spatial/service level	Population served (P)	Sanitation type	Example	Assessment scale
Household	1-10	• VIP latrine, traditional latrine, septic tank	• Usually household housing	1
Cluster	15-25	• 3-5 households on shared septic tank	• Kensington in Lubowa and Naalya	1
	25-50	• 5-10 households on shared eco-san toilet	• Kisumu Nyalenda	1
	50-250	• 10-50 households on a septic tank	• Apartment housing	1
	50-250	• 10-50 household on shared VIP	• CIDI in Makindye	1
Community	500	• Septic tank	• SANA in Manyatta	2
	200-500	• Eco-san toilet	• Kansanga slum, Kampala	2
	800	• Septic tank	• NHCC Namungoona	2
	200-600	• Bio-latrine	• Nyalenda, Bandani, Manyatta	2

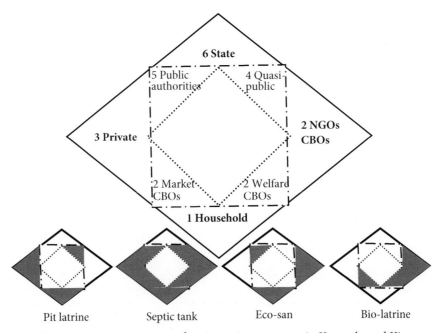

Figure 5.6. Management arrangements of onsite sanitary systems in Kampala and Kisumu.

LVSWSB, provide community sanitation blocks or public toilets that mainly discharge into septic tanks. The operation and maintenance of publicly developed community sanitation blocks are by marketised CBOs (assessed at scale 2); public toilets are offered either by the council (assessed at scale 5), e.g. at Kisumu Bus Park or private (assessed at scale 3) under management contract, e.g. at Kisumu Jomo Kenyatta Grounds. Sanitation in schools and health centres are either on septic tanks or pit latrines and provided by the respective quasi-public administration (assessed at scale 4). Bio-latrines are either provided by NGOs (assessed at scale 2) who solicit for significant percentage of the investment costs, welfare CBOs who are responsible for community mobilisation and contribution, or marketised CBOs who operate the utilities, charge user fees and operation and maintenance. Eco-san in Kampala is provided by KCCA in partnership with NGOs/CBOs (assessed at scale 5), with the latter responsible for operation and maintenance from grant subsidies and little from user fees (assessed at scale 2). Eco-san in Kisumu is promoted by NGOs, earlier on as household facility and currently as community sanitation.

5.7.3 End-user participation: participatory versus technocratic

Participation of end-users in onsite sanitation provision is varied. End-users participate from site selection for facility location and resource mobilisation to operation and maintenance (Table 5.7). From Table 5.7, pit latrines are highly participatory followed by septic tanks, eco-san are technocratic whereas bio-latrines are mixed. Therefore, it can be deduced that not all onsite systems are participatory as expected. However, the low participatory nature of eco-san toilets is attributed to the apparent support by local authorities in Kampala and provisions by NGOs in Kisumu through heavy subsidies and at full costs, respectively. The envisaged high end-user participation in eco-san has been taken over by CBOs at community scale whereas households in Kisumu are only responsible for operation and maintenance.

Table 5.7. Assessment scales of participatory-technocratic dimension in onsite systems.

Sanitation type	Nature of participation	Assessment scale
Pit latrines	• initiation, financing, construction and operation and maintenance	1 & 2
Septic tanks	• initiation, financing and operation and maintenance, with artisan construction in household and private	2
	• resource mobilisation for NGO/CBO supported	3
Eco-san	• operation and maintenance and awareness creation	4 & 5
Bio-latrines	• resource mobilisation	3
	• location and construction	1 & 2
	• operation and maintenance	4

5.7.4 Sanitary flows: separate versus combined water and waste flows

Assessments for the sanitary flow dimension are at scale 1, 2, 4 and 5. Pit latrines are assessed at 2 where the excreta, which comprise of urine and faeces, are dropped into pit latrine without reuse practices. Eco-san, which separates urine and faeces with envisage reuse practices, are assessed at 1 and 2, yellow water and excreta, respectively. There are, however, incomplete reuse chains in eco-san, with the linkage between (peri-)urban farmers and city authorities missing, which is necessary for re-use of stabilised urine and excreta as fertilizer for agriculture and city beautification programmes, respectively. Septic tanks are waterborne onsite sanitation where flows are combined as domestic sewage and thus assessed at 5. Besides, there is flow of faecal sludge to STPs since most septic tanks are emptied by cesspool vacuum trucks, thus assessment at scale 2. Bio-latrines are also waterborne onsite systems, but the water used is small, about 1 litre per flush, no kitchen waste and grey water flow streams, and thus assessment at scale 4, black water. Bio-latrines do tap biogas from valorised wastewater for domestic use.

The high strength characteristics of faecal sludge (Table 5.3), results in shock loads to STPs given that they are overloaded or are not designed to receive faecal sludge. Separate treatment of

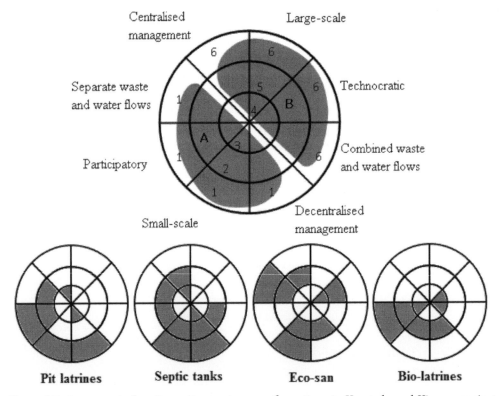

Figure 5.7. Assessment of onsite sanitary systems configurations in Kampala and Kisumu against 4 MM dimensions on a 6 point scale. Areas (A) and (B) represent decentralised and centralised paradigms respectively.

faecal sludge as proposed in Kampala and Kisumu (NWSC, 2004, 2008; LVSWSB, 2005a, 2008) can potentially result in acceptable quality of bio-solids for reuse unlike co-treatment with combined sewage currently. In addition, if anaerobic digestion is applied, biogas will be produced, possibly contributing to local energy supply.

5.8 Conclusion

The assessment, let alone the decision on whether a particular type of onsite sanitation is permanent or transient cannot be judged on one-dimensional criteria. In general, they can be permanent or transient depending not only on population or base flow density, the common criteria used by sanitary agencies, but also on spatial requirements and existing excreta management practices by various actors. Assessments show that bio-latrines are at community level, pit latrines at household level whereas septic tanks and eco-san are operated at both household and community level. In terms of end-user participation, pit latrines and septic tanks can be characterised as participatory, eco-san as technocratic and bio-latrines as mixed. Sanitary flows are separate in pit latrines and eco-san, whereas septic tanks and bio-latrines have domestic and black water flows, respectively. The management arrangements are diverse for all onsite sanitation options: household, private, voluntary sector, quasi-public and state.

Chapter 6.
Assessing the sustainability of sanitary mixtures in Kampala and Kisumu

6.1 Introduction

Sanitation provision in Kampala and Kisumu is a consequence of implementation of different sanitation options. On the one hand, it is through overt public sewer projects and on the other hand, it is a result of search of decent alternative sanitation options outside public provisioning. Presently different stages of sanitation solutions sit next to each other in these cities leading to sanitation mixtures. Pursuing one solution in cities of East Africa which are characterised by spatial, socio-economic, providers and sanitary systems variability is not tenable (Tables 1.1, 1.2, 4.8, 5.4; Figures 2.2, 5.1). Sanitation provisioning in East African cities, therefore, has to be multi-objective and offering multiple options and should as well be judged across technical, spatial, institutional and social aspects of sanitary provision. Therefore, a decision making tool is needed that not only assesses the performance of sanitation mixtures, but also pinpoint elements that need reform, either through improvement or restructuring. Therefore, we propose that sustainability of sanitation mixtures should be assessed based on MM criteria of public and environmental health, accessibility and flexibility. This chapter, therefore, attempts to answer research question 3 of this thesis, to what extent are the existing sanitary systems that are studied in Chapters 3-5 'sustainable'?

One of the techniques often used for assessing infrastructures since 1980s is multi-criteria analysis (MCA) (Van Buuren, 2010). MCA has been widely applied in sanitation infrastructures, especially in assessing sustainability of technological options for implementation (Seghezzo, 2004; Muga & Mihelcic, 2008; Palme, Lundin, Tillman, & Molander, 2005; Palme and Tillman, 2008; Van Buuren, 2010). MCA is used in this study as a tool to assess the extent to which typical sanitation systems in Kampala and Kisumu are sustainable. Utilising MM assessment criteria is not aimed at selecting one preferred option for adoption, but at making choices among imperfect options in the light of local circumstances.

6.2 Methodology

6.2.1 Establishing multi-criteria analysis (MCA)

MCA is a method to make, for a certain situation, the best choice out of several options, where some options score better than others with regard to some objectives, but worse on others (Van Buuren, 2010). MCA is a decision making support tool that aids in making choices among options (Von Münch & Mels, 2008; Hajkowicz & Haggins, 2008). Besides, MCA is an effective tool that not only use ordinal and cardinal data however limited they may be, but also adds structure, audibility, transparency, and rigour to multi-objective decision making (Hajkowicz & Haggins, 2008; Joubert, Stewart, & Eberhard, 2003). MCA, therefore, is considered suitable for the assessment of sustainability of sanitary mixtures, which requires making choices among imperfect options, even with limited data, in coming up with options for specific contexts while pinpointing elements that need improvement or restructuring in order to make them sustainable.

The steps in the MCA are (1) to establish the decision context, (2) to identify the decision options, (3) to establish the assessment criteria and their indicators, (4) to obtain assessment scores from experts, (5) to weigh the criteria, (6) to compute the mean scores and weighted mean scores for each option to derive assessment performance, (7) to examine the results, (8) to assess sensitivity of criteria weight on performance, and (9) to make decision (Hajkowicz & Haggins, 2008; Crown, 2009). Steps 1 and 2 are addressed in Chapter 1 to 5.

The assessment assumptions are (1) that experts are not biased on their assessment towards certain sanitation option, (2) the assessed options are complete systems operating at ideal state, (3) sanitation mixtures are here to stay, (4) modernisation towards sustainability of these sanitation mixtures regarding public and environmental health, accessibility, and flexibility, (5) efficiency of sanitation systems has to improve rather than the capacity, and (6) there is limitation of space, money and willingness.

The decision options to be assessed are the six sanitation systems from Kampala and Kisumu empirical results: conventional urban sewers connected to STPs, satellite sewers with decentralised treatment, and the (household and community) onsite systems, septic tanks, pit latrines, eco-san, and bio-latrines. The sanitation options have been described and provision dimensions configurations graphically presented by means of chart mapping in Chapter 3 to 5. Therefore, this chapter addresses MCA steps 3-9.

6.2.2 Assessment criteria and indicators

Identification of indicators for criteria can be carried out through interviews, stakeholder participation and literature review (Loetscher, 1999; Van Buuren, 2010). In this chapter, indicators are developed through literature review (Table 6.1). MCA requires at least two criteria and two decision options for the assessment task to be done (Hajkowicz & Haggins, 2008). The three MM assessment criteria are public and environmental health, accessibility and flexibility. Public and environmental health, in this context, is concerned with assessing and controlling factors in the urban sanitation environment that can potentially cause public health risks such as uncontrolled spreading of pathogenic organisms and environmental protection risks such as non-controlled spreading of nutrients and non-stabilised organic matter. Accessibility is concerned with access by different urban clientele to different services under different service providers, spaces and social structure. Flexibility is concerned with how technical, institutional and social elements of sanitary provision are resilient, robust and adaptable to changing demands.

6.2.3 Scoring and weighting

MCA is conducted by scoring a finite number of decision options based on a set of criteria and weighing the scores against the same set of criteria to determine scores for each option (Von Münch & Mels, 2008; Hajkowicz & Haggins, 2008). This enables decisions made based on average mean scores (Equation 1) and weighted mean scores (Equation 2 and 3).

$$S_i = \frac{\Sigma S_i}{n} \qquad (1)$$

Where the average mean scores for options i is the sum of expert score on criteria, sub-criteria or criteria S_i divided by the number of experts n.

$$W_i = \frac{\Sigma W}{5n_i} \qquad (2)$$

Where the weighted mean W_i for indicator i is the total weight allocated to criteria W divided by the multiple of the highest rating scale 5 and the number of indicators n.

$$P_{S_i W_i} = s_{i1}w_1 + s_{i2}w_2 + \ldots + s_{in}w_n = \sum_{j=1}^{n} s_{ij}w_i \qquad (3)$$

Where the weighted performance score for option S_i (Crown, 2009), is the average means scores for option i on criterion j is represented by s_{ij}. The weight for each criterion is represented by w_j and total number of criteria by n.

In our assessment, equal weights, i.e. 33.3%, are assigned to the three sets of criteria to avoid bias, but in a later stage, the impact of changing relative weight is further analysed. Equal weighting is considered appropriate because sanitary systems are multi-faceted and multi-objective, whereas individual opinions vary. Six sanitary systems are scored against MM criteria, sub-criteria and indicators through email expert survey. The scores for each sanitary system against each indicator and (sub) criterion are averaged to get the mean scores (Equation 1), which is then multiplied by the weights (Equation 2 and 3). The scoring is done using a regular ranking method based on performance rating scale of 1 to 5, where 1, 2, 3, 4 and 5 stands for very poor, poor, acceptable, good and very good, respectively. The overall performance for each system is calculated by way of linear additive based on the assumption that each indicator and criterion is independent of each other to avoid building uncertainty into the MCA (Equation 3). The results are presented in an evaluation matrix. From the evaluation matrix, analysis is done using average mean scores and weighted mean scores and presented in charts to reduce complexity, aid in discussions and in making judgements. In the assessment, scale 1 & 2 are unsustainable, 3 & 4 are fairly sustainable, whereas scale 5 is sustainable. In terms of intervention measures, scale 1 and 2 require restructuring of systems to make them sustainable, 3 & 4 require some improvements, whereas 5 can be replicated.

The experts were coming from institutions that have a role in sanitary provision (Table 6.2; Appendix 2). Two persons were selected from public agencies: NWSC who are mandated to provide urban sewerage and KCCA who are mandated to provide and regulate alternative sanitation in Kampala. In the private sector, an expert from an engineering company that prepared the 2008 sewerage plan for Kisumu, besides implementing sanitary projects across East Africa, was surveyed. In academia, three persons were sampled, two environmental engineering lecturers with track records in sanitary consultancy services and a post-doctoral researcher with technical background and working on multidisciplinary sanitation projects.

From the expert survey assessment results data sheet, urban, septic tank and pit latrine systems were scored by all 6 experts; satellite and bio-latrine systems were scored by 5 experts; whereas eco-san system was scored by 4 experts.

If performance rating is 1 (very poor), 2 (poor), 3 (acceptable), 4 (good) and 5 (very good), then systems with score (S) of 1-2 are unsustainable, 3-4 fairly sustainable and 5 sustainable.

Table 6.1. Modernised mixtures assessment criteria, sub-criteria and indicators.

Criteria	Sub-criteria	Indicators	Reference
Public and environmental health	Carrying capacity	• system can accommodate high base-flow density	NWSC (2004)); Veenstra (1996))
		• system can accommodate high population density	Fang (1999); Sinnatamby (1983); Mara (2008); Ho (2005)
	Low emissions	• low malodorous	Muga & Mihelcic (2008)
		• low methane emission to air	Seghezzo (2004); Guest, Daigger, Corbett, & Love (2010); Palme *et al.* (2005)
		• low emission to soil, surface and groundwater	Seghezzo (2004); Guest *et al.* (2010); Palme *et al.* (2005
	Reduced exposure to hazards	• reduced exposure to downstream users	Zurbrugg & Tilley (2007); Van Buuren (2010)
		• reduced exposure to re-users	Palme *et al.* (2005); Van Buuren (2010)
		• reduced exposure to users	Palme *et al.* (2005); Van Buuren (2010)
		• reduced exposure to workers	Palme *et al.* (2005); Van Buuren (2010)
	Removal efficiency	• high organic load removal	Seghezzo (2004); Van Buuren (2010)
		• high pathogen removal	Seghezzo (2004); Vanish and Shah (2008)
	Resource conservation	• low energy consumption	Van Lier & Lettinga (1999); Tsagarakis, Mara, & Angelakis (2003); Palme *et al.* (2005)
		• low water use	Van Lier & Lettinga (1999)
		• recycling & reuse of organic matter, nutrients & water	Seghezzo (2004); Vashi & Shah (2008)
Accessibility	Institutional accessibility	• high flexibility in service provision levels	Palme (2009)
		• low level expertise in design to operation	Zurbrugg & Tilley (2007); Van der Vleuten-Balkema (2003)
		• low requirement for subsidy/cross subsidy	Kessides (2004)
	Physical accessibility	• insensitive to settlement type	Oosteveer & Spaargaren (2010); Mara (2008)
		• low land use requirement	Seghezzo (2004); Tsagarakis *et al.* (2002)
		• low requirement for mandatory distance to servicing	Milman & Short (2008)
	Social accessibility	• flexible or low payment for facilities or services	Kessides (2004)
		• high convenience	Zurbruegg & Tilley (2007); Van Buuren (2010)
		• low per capita construction and operation costs	Seghezzo (2004); Tsagarakis *et al.* (2002); Guest *et al.* (2010)
		• low requirement for service agreements	Kessides (2004)

Criteria	Sub-criteria	Indicators	Reference
Flexibility	Institutional flexibility	• low requirement to institutional support	Seghezzo (2004); Guest *et al.* (2010); Vashi & Shah (2008)
		• simplicity of procedures	Zurbruegg & Tilley (2007); Van Buuren (2010)
	Social flexibility	• consider issues of women, children, elderly, disabled	Van Buuren (2010); Zurbruegg & Tilley (2007)
		• flexible to political shocks	Dunmande (2002); Oosteveer & Spaargaren (2010)
		• low requirement for end user awareness	Guest *et al.* (2010); Zurbruegg &Tilley (2007)
	Technical flexibility	• availability of appropriate labour locally	Muga & Mihelcic (2008)
		• easily adaptable to new conditions and requirements	Dunmande (2002); Seghezzo (2004); Guest *et al.* (2010)
		• flexible in planning and construction standards	Guest *et al.* (2010)
		• independent of external suppliers	Van Lier & Lettinga (1999); Van Buuren (2010)
		• low sensitivity to irregular maintenance	Seghezzo (2004); Massoud *et al.* (2009)

Performance rating scores to the indicators are (S): 1 very poor; 2 poor; 3 acceptable; 4 good, 5 very good.

Table 6.2. Expert survey decision makers' panel.

No.	Institutional affiliation[1]	Designation of the expert	Country
1	NWSC	chief analyst	Uganda
1	KCCA	sanitary engineer	Uganda
1	Odongo and Odongo company	senior environmental engineer	Kenya
2	Ardhi University of Dar es Salaam	environmental engineering lecturer	Tanzania
1	PROVIDE project	post-doctoral researcher	Netherlands

[1] NWSC: National Water and Sewerage Corporation; KCCA: Kampala Capital City Authority; PROVIDE: partnership research on viable environmental infrastructure in East Africa.

6.3 Results and discussions

6.3.1 Mean score performance assessment

The summary of performance assessment of criteria, sub-criteria and indicators is presented in the assessment matrix (Table 6.3). The mean scores from the expert survey, with deviation from the mean score, are presented in Figure 6.1. The overall standard deviation per sanitary system is urban (1.5), satellite (1.4), septic tank (1.2), pit latrine (1.5), eco-san (1.3) and bio-latrine (1.1). The overall performances on MM criteria from the mean scores (Figure 6.1) show that all the six sanitary systems are rated acceptable.

The performance per MM criterion (Figure 6.2) shows that urban, satellite and bio-latrines are rated good in public and environmental health and acceptable in flexibility, with urban and satellite rated poor and acceptable in accessibility respectively. Septic tanks and eco-san are rated acceptable in both public and environmental health and accessibility whereas in flexibility, they are rated as good. Pit latrines are rated poor in public and environmental health, good in accessibility and very good in flexibility.

The mean score performance data per sub-criterion (Figure 6.3) shows that sewerage based sanitary systems, urban and satellite, are rated good in carrying capacity and removal efficiency; poor in resource conservation, technical flexibility and institutional flexibility; and acceptable in

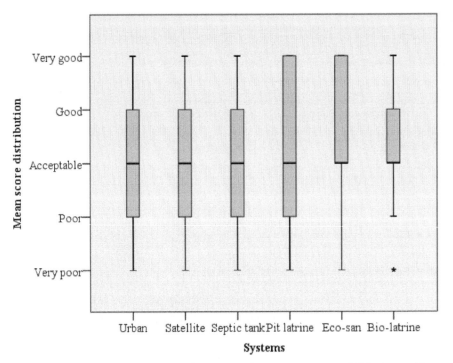

Figure 6.1. Distribution and mean score performances of sanitary systems on MM criteria. *Assessment scores outside the boxplots range.

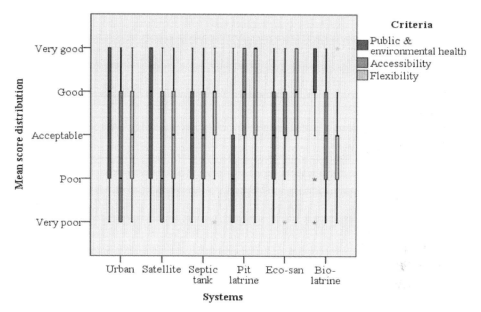

Figure 6.2. Distribution and mean score performance of sanitary systems on MM criterion.
*Assessment scores outside the boxplots range.

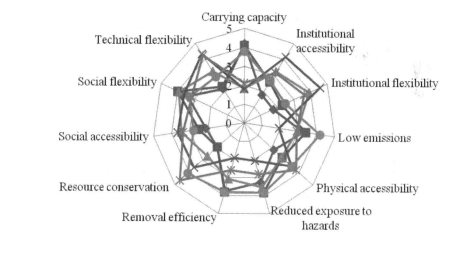

Figure 6.3. Mean score performance of sanitary systems per MM sub-criterion.
Performance rating: 1 = very poor; 2 = poor; 3 = acceptable; 4 = good, 5 = very good.

Table 6.3. Performance matrix of typical sanitary systems in Kampala and Kisumu.[1]

(Sub)Criteria	Indicators	Weight	Urban		Satellite		Septic tank		Pit latrine		Eco-san		Bio-latrine	
		W	S	W*S	S	W*S	S	W*S	S	W*S	S	W*S	S	W*S
Accessibility														
Institutional accessibility	high flexibility in service provision levels	0.67	2.5	1.7	4.0	2.7	3.3	2.2	3.0	2.0	3.3	2.2	3.2	2.1
	low level of expertise	0.67	1.3	0.9	1.8	1.2	2.8	1.9	4.7	3.1	2.5	1.7	2.0	1.3
	low requirement for subsidy/cross subsidy	0.67	1.5	1.0	2.0	1.3	3.5	2.3	4.8	3.2	3.8	2.5	2.8	1.9
Physical accessibility	insensitive to settlement type	0.67	1.5	1.0	2.0	1.3	1.8	1.2	2.8	1.8	3.3	2.2	3.3	2.2
	low land use requirement	0.67	2.3	1.6	2.8	1.9	3.3	2.2	4.0	2.7	4.3	2.8	4.2	2.8
	low requirement for distance to servicing	0.67	2.5	1.7	3.7	2.5	2.8	1.8	3.4	2.3	3.7	2.5	2.8	1.8
Social accessibility	flexible/low payments for facilities/services	0.67	1.7	1.1	2.0	1.3	3.3	2.2	4.2	2.8	3.8	2.5	3.2	2.1
	high convenience	0.67	4.3	2.9	4.6	3.1	4.2	2.8	2.2	1.5	2.0	1.3	2.8	1.9
	low per capita construction & operation	0.67	1.3	0.9	1.8	1.2	2.7	1.8	3.8	2.6	3.5	2.3	2.8	1.9
	low requirement for service agreements	0.67	1.2	0.8	1.6	1.1	3.7	2.5	4.5	3.0	4.3	2.8	2.4	1.6
Sub-total equal weighted performance		33.3		13.5		17.6		21.0		25.0		22.9		19.7
Flexibility														
Institutional flexibility	low requirement to institutional support	0.67	1.2	0.8	2.0	1.3	4.2	2.8	4.3	2.9	4.3	2.8	2.2	1.5
	simplicity of procedures	0.67	2.3	1.6	2.6	1.7	3.7	2.5	4.8	3.2	3.0	2.0	2.8	1.9
Social flexibility	consider women, children, elderly, disabled	0.67	3.5	2.3	4.3	2.9	3.3	2.2	1.8	1.2	1.7	1.1	3.0	2.0
	flexible to political shocks	0.67	3.2	2.1	4.2	2.8	4.2	2.8	4.8	3.2	5.0	3.4	3.8	2.5
	low requirement for end user awareness	0.67	3.5	2.3	3.6	2.4	4.2	2.8	4.2	2.8	2.5	1.7	3.0	2.0
Technical flexibility	availability of appropriate labour locally	0.67	2.0	1.3	2.6	1.7	3.5	2.3	4.7	3.1	4.5	3.0	3.0	2.0
	easily adaptable to new conditions	0.67	2.8	1.9	1.6	1.1	3.2	2.1	4.2	2.8	4.5	3.0	3.2	2.1
	flexible in planning/ construction standards	0.67	2.5	1.7	3.2	2.1	2.3	1.6	4.0	2.7	4.3	2.8	3.6	2.4
	independent of external supplier	0.67	1.8	1.2	2.0	1.3	3.0	2.0	4.8	3.2	4.3	2.8	3.0	2.0
	low sensitivity to irregular maintenance	0.67	1.2	0.8	1.6	1.1	3.8	2.6	3.8	2.6	3.5	2.3	1.6	1.1
Sub-total equal weighted performance		33.3		16.1		18.6		23.6		27.8		25.1		19.6

Public & Environmental health

		Weight												
Carrying capacity	system can accommodate high m³/ha*d	0.47	4.2	2.0	4.2	2.0	2.0	0.9	2.0	0.9	2.0	0.9	3.6	1.7
	system can accommodate high P/ha	0.47	4.2	2.0	4.0	1.9	1.7	0.8	1.7	0.8	2.3	1.1	4.0	1.9
Low emissions	Low malodorous	0.47	3.7	1.7	4.0	1.9	3.3	1.6	2.0	0.9	2.8	1.3	3.8	1.8
	low methane emission	0.47	2.6	1.2	2.5	1.2	2.8	1.3	2.2	1.0	2.8	1.3	4.6	2.2
	low pollution-soil, surface & groundwater	0.47	3.5	1.6	3.6	1.7	2.7	1.3	1.7	0.8	3.3	1.5	4.2	2.0
Reduced exposure to hazards	reduced exposure to downstream users	0.47	2.6	1.2	2.8	1.3	3.3	1.6	2.2	1.0	3.3	1.5	3.8	1.8
	reduced exposure to re-users	0.47	3.2	1.5	3.8	1.8	3.2	1.5	1.8	0.9	2.8	1.3	4.0	1.9
	reduced exposure to users	0.47	4.0	1.9	4.6	2.2	4.2	2.0	2.7	1.3	3.0	1.4	4.0	1.9
	reduced exposure to workers	0.47	3.2	1.5	4.0	1.9	3.5	1.6	1.7	0.8	2.3	1.1	3.6	1.7
Removal efficiency	high organic load removal-BOD, COD, TSS	0.47	4.0	1.9	4.0	1.9	3.5	1.6	2.2	1.0	2.3	1.1	4.2	2.0
	high pathogen removal	0.47	3.5	1.6	3.6	1.7	2.7	1.3	1.7	0.8	3.0	1.4	3.4	1.6
Resource conservation	low energy consumption	0.47	1.7	0.8	2.2	1.0	4.2	2.0	4.8	2.3	4.8	2.2	4.4	2.1
	low water use	0.47	1.5	0.7	1.4	0.7	2.2	1.0	4.5	2.1	4.8	2.2	3.6	1.7
	recycle/reuse-organic, nutrients, water	0.47	2.7	1.3	2.2	1.0	1.7	0.8	1.5	0.7	4.5	2.1	4.2	2.0
Sub-total equal weighted performance		33.3		20.9		22.0		19.2		15.3		20.4		26.0
Total equal weighted performance			**53.8**		**60.9**		**67.2**		**71.4**		**71.8**		**65.3**	

[1] The scores in **bold and italic** indicate the elements that are considered sustainable in a sanitation system; the scores in un-shaded cells indicate the elements that require improvements in order to make them sustainable; the scores in shaded cells indicate elements that need restructuring in order to make them sustainable.

low emissions. In reduced exposure to hazards and social flexibility, satellite systems are rated as good whereas urban systems are rated as acceptable. Septic tanks are rated as poor in carrying capacity and good in reduced exposure to hazards and institutional and social flexibility whereas in low emissions and physical accessibility, they are rated as acceptable. Pit latrines are rated poor in carrying capacity, low emissions and reduced exposure to hazards whereas in institutional flexibility, they are rated as very good. They are also rated as good in institutional flexibility and accessibility, removal efficiency, resource conservation, social accessibility and flexibility. Eco-san latrines are rated poor in carrying capacity, good in institutional and technical flexibility and physical accessibility, and very good in resource conservation. Bio-latrines are rated as acceptable to good in sub-criteria, with good in carrying capacity, low emissions, removal efficiency, resource conservation, and reduced exposure to hazards.

6.3.2 Weighted performance assessment

The weighted performance of sanitary systems per criterion (Figure 6.4) show that in accessibility criteria, urban systems are rated as poor; satellite, septic tank and bio-latrine are rated as acceptable; and pit latrine and eco-san systems are rated as good. In public and environmental health criterion, urban, septic tank and pit latrine systems are rated as acceptable; whereas satellite, eco-san and bio-latrine are rated as good. In flexibility criteria, the performance rating in urban, satellite and bio-latrine systems are acceptable; whereas septic tank, pit latrine and eco-san systems are good. The reference performance is 33.3%, which is the maximum weighted performance a criterion can attain.

The overall weighted performance scores on MM criteria are presented in Figure 6.5. From the figure, urban and satellite systems are rated acceptable; whereas the remaining four systems are rated good.

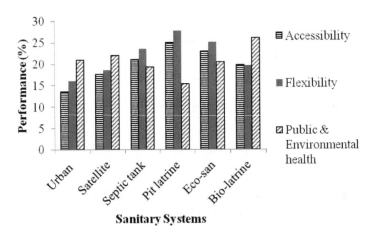

Figure 6.4. Weighted performance scores of sanitary systems per MM criterion.
Performance rating: 1-7 = very poor; 7-14 = poor; 14-21 = acceptable; 11-28 = good; 28-34 = very good.

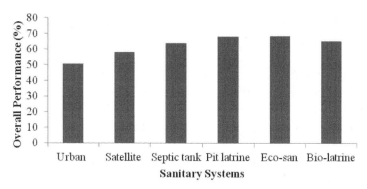

Figure 6.5. Weighted performance scores of sanitary systems on MM criteria.
Performance rating: 1-20 = very poor; 20-40 = poor; 40-60 = acceptable; 60-80 = good; 80-100 = very good.

6.3.3 Impact of criteria weight on overall performance

The sensitivity of criteria weight is assessed by varying criteria weights systematically by doubling to test the robustness of assessment results. Variation of weights helps in determining the relative impacts of the weights on the overall score of the sanitary system. A further analysis of the criteria weight distribution is imperative, since the surveyed experts were not asked to give preference weights. The criterion weight being assessed was increased from 33.3 to 66.6% and the others decreased from 33.3% to 16.7%, such that the total weight remains 100%. The results of this analysis on criteria weight sensitivity are presented in Figure 6.6. Doubling the weight of accessibility increases overall performance of pit latrine, but decreases the performances of urban, satellite and bio-latrines. Doubling the weight of public and environmental health increases the performance of urban, satellite and bio-latrine systems, but decreases the performance of septic tank, pit latrine and eco-san. Doubling the weight of flexibility increases the performance of septic tank, pit latrine and eco-san, but decreases the performance of urban, satellite and bio-latrine.

Results presented in Figure 6.6 indicate that although the performances vary with increases in criterion weight, changes in performance range between 3% in septic tanks to 15% in pit latrines, but with no performance shift across rating scales. Generally, increase in weights have impact on systems that already have high performance with respect to that criterion whereas systems that have low performance, the impact is small or even negative.

6.3.4 Making choices among imperfect options

There is no system in the overall performance that is rated very good; neither is there a system that is rated very poor, but generally all systems are rated acceptable to good. Therefore, there is no system that scores high in all aspects of modernised mixtures sustainability criteria. This means that there are no bad sanitation solutions, but they are context specific, thus accepting the existence of sanitary mixtures. Nevertheless, the results can inform decision making. First, we can pinpoint the elements in each sanitary system that can inform different intervention measures.

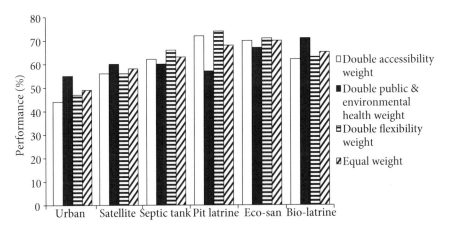

Sanitary Systems

Figure 6.6. Impact of doubling criterion weights on overall performance in reference to equal weighting. Performance rating: 1-20 = very poor; 20-40 = poor; 40-60 = acceptable; 60-80 = good; 80-100 = very good.

For instance, from indicator performance scores (Table 6.3) we can deduce that most elements in the current sanitary systems need improvements, a few need restructuring, and a handful of the elements are considered sustainable. Second, we can select sanitary options based on performance. For instance, if we use Figure 6.3 to select sanitary options, it results in different sanitary systems (Table 6.4), which can inform different policy and local contexts.

Third, we can inform policy decisions. For instance, from the above analysis on weight distribution sensitivity, we can deduce that policies aimed at stringent environmental discharge requirements will have a positive impact on urban, satellite and bio-latrine systems, but will impact negatively on systems that have high accessibility and flexibility, e.g. septic tanks and pit latrines.

Table 6.4. Selection of sanitary options from sub-criteria mean score performances.

Sub-criteria	Sanitary option
High population and base flow density	urban, satellite and bio-latrine
Reduced exposure to hazards	satellite, bio-latrine and septic tanks
High removal efficiency	urban, satellite and bio-latrines
High physical accessibility	eco-san
High resource conservation	eco-san, bio-latrine and pit latrine
High social accessibility	pit latrine
High social flexibility	satellite, septic tank and pit latrine
High institutional flexibility	pit latrine, septic tank and eco-san

Therefore, programmes for sanitation intervention can go for improvement options where systems already have higher performance; whereas those with lower performances may need comprehensive or even systems reconfigurations for significant impacts to be realised.

6.4 Conclusion

Assessment of sanitary systems shows that there is no system that is outcompeted in the overall sustainability performance as they are all assessed as acceptable. Assessment per criterion, however, shows that urban and satellite systems are assessed as good in public and environmental health, poor in accessibility and acceptable in flexibility. Bio-latrines are assessed as good in public and environmental health and acceptable in accessibility and flexibility. Pit latrines are assessed as very good in flexibility, good in accessibility and very poor in public and environmental health. Eco-san systems are assessed as good in flexibility and acceptable in accessibility and public and environmental health. Septic tanks are assessed as good in flexibility and acceptable in accessibility and public and environmental health. Sanitary system choices, consequently, are made among imperfect options, which call for balancing the MM criteria and elements to suit different policy contexts. Besides, based on indicators, some elements in sanitary systems can be considered sustainable; some need improvements whereas others need restructuring.

The high number of sub-criteria (11) and indicators (34) used depict the complexity of sanitary provision. The omission of weights during the survey was meant to avoid bias among experts; whereas the equal weights assigned ignored the importance of each criterion, but this was taken care of by analysing the sensitivity of this presumption. Scoring on scale 1-5 seems to be appropriate since no quantitative data were available or derived; besides, it was used to reduce complexity. The MCA has proven to be a useful tool in assessing sustainability of sanitary mixtures even in the context of limited data availability. The existence of sanitary mixtures means multiple possibilities, which coupled with the three MM assessment criteria, meets the MCA requirement. Any new sanitary option or approach may not be scored by all experts or if they do, it may be scored low in the MCA owing to low awareness of such system possibilities. Besides, getting all-round experts to participate in the performance assessment is not easy.

Chapter 7.
Conclusions, reflection and further outlook on modernised mixtures approach

7.1 Introduction

This thesis aims to contribute to developing alternative views on the assessment and provision of sanitary systems taking East African cities of Kampala and Kisumu as case studies. The results of survey and empirical exploration show that sanitary provision in Kampala and Kisumu consists of mixtures. These mixtures, however, are not sustainable and thus there is a need for an intervention strategy. This thesis postulates that the daunting sanitation challenges in East African cities can be effectively addressed by a novel assessment tool, denominated as the modernised mixtures (MM) approach, which characterises the existing sanitary mixtures and elucidates shortcomings for improvement. To assess the potentials of the MM approach in East African cities, four research questions were formulated to structure the research, and which also structure this concluding chapter:

1. What are the types of sanitary systems in Kampala and Kisumu?
2. What are the configurations of sanitary systems in terms of MM dimensions in Kampala and Kisumu?
3. To what extent are the existing sanitary systems in Kampala and Kisumu considered sustainable following MM sustainability criteria?
4. To what extent does a MM approach provide a useful conceptual model and tool for assessing, prescribing and generalising on sanitary systems in East African cities?

This conclusive chapter addresses the research questions in a comprehensive way and generalises on the MM approach as positioned in Chapter 2 using the empirical findings of Chapter 3 to 6.

7.2 Spatial-technical characteristics of sanitary systems

This section presents the answers on part of research Question 1 and 2, what are the types and configurations of sanitary systems in Kampala and Kisumu?

Empirical findings in Chapter 3 to 5 show that there are different types of sanitary systems in Kampala and Kisumu with different scales and servicing different spatial structures (Tables 7.1, 5.6, 4.5, 3.10; Figure 7.2).

From Table 7.1, it can be deduced that:
- the planned city core, e.g. central business districts, civic centre, planned neighbourhoods and old industrial areas are serviced by urban and septic tank systems;
- planned peri-urban areas are serviced by urban, satellite, septic tank, traditional pit (TP) latrine and ventilated improved pit (VIP) latrine systems;
- unplanned peri-urban areas are serviced mostly by shared septic tank, VIP latrine and bio-latrine systems;
- rural areas are serviced by TP latrines, VIP latrines, eco-san toilet and satellite systems, with the latter for planned areas.

Table 7.1. Spatial-technical structure of sanitary systems in Kampala and Kisumu.[1]

Scale/size (P)	Sanitary system	Urban Planned	Peri-urban Planned	Peri-urban Unplanned	Rural Planned	Rural Unplanned
Household (5-50)	TP latrine					
	VIP latrine					
	Septic tank					
	Eco-san					
Cluster (10-200)	Septic tank					
	VIP latrine					
Community (50-1,500)	Septic tank					
	Eco-san					
	Bio-latrine					
	Satellite					
(1,500-5,000)	Satellite					
	Urban					
Small urban (5,000-50,000)	Satellite					
	Urban					
Medium urban (50,000-250,000)	Satellite					
	Urban					
Large urban (>250,000)	Urban					

[1] The shaded cells indicate the sanitary systems that are applicable for the respective spatial-technical structures.

Generally, urban systems can be characterised as large as they are operated at medium-urban scale; satellite systems are intermediate as they serve community, neighbourhood and small-urban scales; whereas onsite systems, e.g. septic tanks, eco-san, bio-latrines and pit latrines can be considered small scale as they serve at household, cluster or community scale.

Urban systems (Chapter 3) can be considered as conventional and centralised since they are conventionally designed and constructed, are supply driven, require large contiguous areas to form economical service units, supported by large sewage treatment plants (STPs) and based on high water consumption, e.g. domestic water supply of 115-250 l/ca*d for residential, 15-20 m³/ha*d for non-residential, 17.5 m³/ha*d for industrial, and 10 m³/ha*d for commercial and institutional sectors(MWI, 2005, 2008b; LVSWSB, 2005a, 2008; NWSC, 2004, 2008). Sewerage areas are based on attainment of base flow density of 10 m³/ha*d and population density of 200 P/ha in Kampala and 120 P/ha in Kenya (MWI, 2008b; NWSC, 2004). They targets planned areas, settlements occupied by middle and high-income groups and strategic and sensitive areas such as civic, commercial and industrial centres and government installations. Unplanned settlements are either dismally sewered or not sewered despite attaining the density thresholds, even in situations where trunk sewer lines passes through them. Satellite systems (Chapter 4) can be considered to be intermediate semi-collective systems between urban and household/cluster systems (Table 7.1; Figure 7.2). Satellite systems utilise gravity sewers and they are equipped with localised STPs (Table 4.3). They are based on conventional sewerage design and construction protocols developed for large-scale urban systems, e.g. water consumption and planning, design and construction standards. They service planned middle and high-income residential settlements, industrial complexes, endowed public and private universities and government facilities, e.g. prison, police lines and institutes (Table 4.2). Satellite systems, however, are based on fixed population, area and user group. Consequently, they are not flexible to population growth and territorial expansion, thus are exclusionary, with those in satellite areas enjoying sewer services and those around excluded. Onsite sanitation systems (Chapter 5) can be considered as decentralised as they are demand driven, site and user specific, simple and non-networked systems applied at planned and unplanned settlements, household, cluster and community level (Tables 5.6 and 7.1). They are often perceived as transient solutions in cities but can be a permanent solution if population and base flow densities are low or if they are located in suitable areas, accessible and linked to efficient and sanitary emptying practices.

Storm water in urban systems is separated from sewage flows, both in design and practice. Sewage flows are mixed with industrial wastewater and co-treated with faecal sludge in centralised systems whereas in catchment systems, sewage is co-treated with faecal sludge. Due to cross connection via inspection chambers, faulty manholes and infiltration, storm water ends up mixing with sewage flows. For instance in Kampala's Bugolobi STPs, storm water accounts for 35% of flows (NWSC, 2008). Reuse and resource recovery practices are done through the sale of biosolids for agricultural use and informal irrigation of vegetables with sewage effluent and fishing in maturation ponds. Effluent discharge standards are stringent, contradictory and unrealistic. Concentrations of NH_4^+-N often equal those of TN, whereas phosporous removal requires expensive technologies for which financing is likely a constraint. In satellite systems, storm water and sewage flows are separated in design and practice. Storm water is drained by open drains and sewage by closed drains. The sewage flows are land use or facility specific: industrial, domestic or institutional. So far, there are no reuse and resource recovery practices, except where maturation ponds are used informally as fish farms. The sewage collected locally is treated locally close to the point of generation without co-treatment with faecal sludge from onsite sanitation. Onsite sanitary systems can be categorised in terms of flows as domestic water for septic tanks and black water for bio-latrines, excreta in pit latrines and eco-san and urine diversion in eco-san toilets. Deliberate source separation is hardly applied, except in eco-san piloting projects that demonstrate reuse of stabilised excreta and bio-latrines for tapping and use of biogas as energy source. Obviously, when applying latrine sanitation, black water is source separated from the grey water streams.

7.3 Institutional characteristics of sanitary systems

This section attempts to answer parts of research Question 1 & 2, focusing on the management of sanitary systems and level of end-user participation.

Traditionally, from 1900s until 1972 in Kampala and 1900s until 2003 in Kisumu, local authorities (LAs) provided water supply, sewerage, drainage, solid waste management and faecal sludge management. Services, thus, were centralised at the local authority (LAs), offering integrated planning, coordination and control. Currently sanitary services provisions are the responsibility of multiple providers (Table 7.2; Figure 7.1).

The institutional framework for sewerage is centralised at the national level and hierarchical in nature (Figure 3.8), with central government controlling the entire sector through the Minister. Urban sanitary systems are provided by public sewerage authorities under new public management (NPM) arrangements (Table 3.9). Asset ownership is seperated from operation and maintenance, where the former is done by a public corporation and the latter by a public service provider, resulting in public-public form of management. They operate with performance contracts. Asset ownership is separated from utility operation, which can be done by private sector parties through outsourcing of services such as design, construction, and rehabilitation. Satellite systems are developed and managed by housing companies, quasi-public institutions (universities) and the state. Only in Ntinda there is a neighbourhood association that operates the utility (Figure 7.1). The operators are not appointed as sewerage authorities as required in the regulations, and therefore, they do not operate on performance contracts.

Onsite sanitation systems are regulated by the Ministry of Health, with LAs being responsible for enforcement. They are developed and managed by a number of stakeholders: LAs for group or pilot projects, households for owner-occupiers, landlords for rental premises, CBOs and NGOs for group sanitations, and private entrepreneurs for private toilets. In pilot projects for group

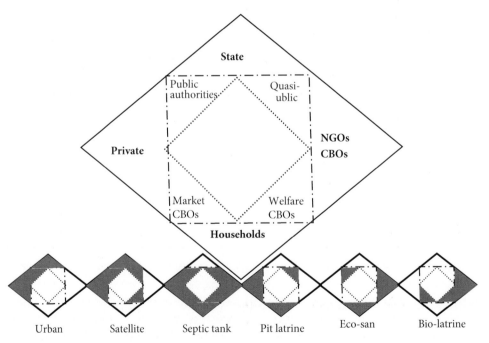

Figure 7.1. Management arrangements found in Kampala and Kisumu sanitary systems. Shadings indicate the management arrangement(s) that apply for a specific system.

Table 7.2. Sanitary services provision institutional arrangement in Kampala and Kisumu.[1]

System type	Developer	Ownership	Operator	Form of provision	Form of management	Management instrument
Urban sanitary systems (Kampala and Kisumu)						
Kampala urban	KCCA NWSC	NWSC	KWP	public sewer with standard monthly payment	public-public	IDAMCs ZPCs
Kisumu urban	MCK LVSWSB	MCK & LVSWSB	KIWASCO		public-public	SPA ALA
Satellite sanitary systems (Kampala)						
Bugolobi	NHCC	NHCC	NHCC	private sewer	self	absent
Kyambogo	university	university	university	private sewer	self	absent
Mukono	university	university	university	private sewer	self	absent
Unise	state	institute	institute	private sewer	self	absent
Naalya	NHCC	NHCC	NHCC	private sewer	self	absent
Namboole	state	state	SMC	private sewer	public-public	absent
Ntinda	NHCC	NHCC	association	private sewer	private-CBO	absent
Onsite sanitary systems (Kampala and Kisumu)						
TP latrine	household	household	household	private toilet	self	absent
	landlord	landlord	households	private toilet	private-user	absent
Lined VIP	LAs/SAs	users	users	shared toilet	public-users	absent
Septic tank	household	household	household	private toilet	self	absent
	private	enterprise	private	private pay toilet	private enterprise	absent
	landlord	landlord	private	private toilet	private-users	absent
	NGOs/CBOs	community	CBOs	community pay toilet	NGO/CBO-community	agreements
	LAs	public	LAs	public pay toilet	self	absent
			private	public pay toilet	public-private	contracts
			CBOs	public pay toilet	public-CBO	contracts
Bio-latrine	NGOs/CBOs	community	CBOs	community pay toilet	NGO-CBO-community	agreements
Eco-san	NGO	household	households	private toilet	NGO-user	absent
	LAs	community	CBOs	shared toilets	public-CBOs	agreements

[1] ALA: Asset Lease Agreement; IDAMC: Internally Delegated Area Management Contract; ZPCs: Zonal performance contracts; KCCA: Kampala Capital City Authority; NWSC: National Water and Sewerage Corporation; MCK: Municipal Council of Kisumu; LVSWSB: Lake Victoria South Water Services Board; NHCC: National Housing and Construction Company; LAs: Local Authorities; SAs: Sewerage Authorities; NGO: Non-Governmental Organisation; CBOs: Community-Based Organisations; KWP: Kampala Water Partnership; KIWASCO: Kisumu Water and Sewerage Company; SPA: Service Provision Agreement; SMC: Sports Management Council.

sanitation there are partnerships between LAs and NGOs/CBOs (Figure 7.1, Table 5.4). The forms of partnerships in sanitary provision are public-public, private-CBO, private-users, NGO/CBO-community, public-CBO, NGO-users (Table 7.2). The forms of utility provision are diverse (Table 7.2; Figure 7.1): there are public sewers, private sewers, private toilets, private pay toilets, public pay toilets, community pay toilets and shared toilets.

The assessments above (Table 7.2; Figure 7.1), which are summaries of empirical findings (Tables 4.7 and 5.4; Figures 3.8, 3.10, 4.6 and 5.6) indicate that management arrangements are diverse:

- private arrangements are visible in the provision and management of most onsite systems, except eco-san and bio-latrines, which are new sanitary options;
- voluntary sector arrangements are visible in the provision and management of septic tanks, eco-san and bio-latrines;
- public authorities are visible in the development and management of urban, satellite and septic tanks;
- households provide and manage septic tanks and pit latrines;
- quasi-public institutions provide satellite, septic tanks and pit latrine systems;
- marketised CBOs provide and manage septic tanks and eco-sans whereas welfare CBOs provide and manage bio-latrines, septic tanks and satellite systems.

Participation of end-users in urban systems, e.g. households and community groups is limited to participation in occasional surveys and sensitisation programmes. End-users are, in most cases; participate in payment of bills and charges and reporting of sewage overflows and blockages. Consequently, urban systems are technocratic, expert driven and public utility centred. In satellite systems, end-users do not participate in planning, design and construction, but much less so in operation and maintenance of STPs. A framework for participation of households, communities or NGOs is absent. Therefore, satellite systems can be characterised as technocratic as they have low end-user participation, and apply standards, technologies and construction protocols developed for large-scale systems. Households or communities are not given choices and are captive consumers. In onsite systems, end-users are involved in development, operation and maintenance (Tables 5.4 and 5.8), and thus can be considered as participatory. The constructions are made by local artisans, with or without expert design or LAs approval. The siting of facilities is mostly done with or by the targeted households or communities.

7.4 Configurations of sanitary systems

This section answers research Question 2: what are the configurations of existing sanitary systems in terms of MM dimensions in Kampala and Kisumu?

Combining the mapped sanitary configurations dimensions in Kampala and Kisumu from Figures 3.11, 4.7 and 5.7, and Tables 4.5 and 4.7, it results in configurations shown in Figure 7.2. The figure 7.2 reveals that there are three categories of sanitary configurations: conventional, traditional and mixed. Urban sanitary systems can be assessed as conventional because they are centralised, large-scale, under technocratic public management that combine water supply and sewage management and offer standardised services. Onsite systems generally can be assessed to

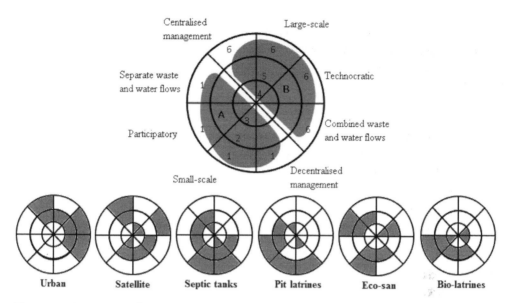

Figure 7.2. Assessment of sanitary systems configurations in Kampala and Kisumu against 4 MM dimensions on a 6 point scale. Areas (A) and (B) represent decentralised and centralised paradigms respectively. They are summaries of findings in Chapters 3-5.

be traditional as they are highly decentralised at individual site, flexible, participatory, and applied at household, cluster or community scales. Besides, they require the involvement of end-users with artisanal skills and multiple local providers. Besides, water and wastewater are not combined, so excreta are flowing to STPs via cesspool exhausters or to the environment via manual emptying.

The mixed (hybrid) systems are satellite or community scale systems. *Satellite systems* are intermediate systems between urban and household onsite, applying conventional sewer standards and construction protocols. End user involvement is mostly lacking. Satellite systems are developed by various service providers who are not sewerage authorities. Water supply and sewage are separately managed. *Community-scale systems*, the bio-latrines, are developed as bio-centres, which are planned as multifunctional community utility with toilet, shower, water vending, and rental units. They capture biogas for use as energy source, are designed by experts, but constructed by artisans and local community members. They also have the same accessibility distance of about 60 m which is similar to sewerage systems. *Community septic tanks* are mixed configurations as the connected apartments are fitted with conventionally designed sewers that discharge into septic tanks as transition to future satellite STPs. Besides, the infrastructures are designed and constructed by sewerage experts on the basis of a defined population size and hydraulic flows. So far settled sewers or small bore sewers connecting septic tank outflows for off-site treatment and/or discharge have not been installed.

The different sanitary configurations are attributed to the existence of differentiated spatial structure, providers and service levels inherent in Kampala and Kisumu cities (Tables 7.1, 7.2, 3.7; Figures 7.1, 3.7). Sanitary mixtures constitute a multiplicity of solutions exhibiting different configurations. Different sanitary configurations require different spaces and institutional arrangements in order to fit the local conditions (Tables 7.1 and 7.2).

7.5 Sustainability of sanitary mixtures

This section answers research Question 3, to what extent are existing sanitary systems in Kampala and Kisumu sustainable?

The performances of the six sanitary systems (Table 6.3; Figures 6.1-6.4) show that there is no system that completely outperforms the others. For instance in the multi-criteria analysis (MCA) as performed and presented in Chapter 6, no system is rated very good in performance, but they are generally rated acceptable to good. Varying the assigned relative weight of the various criteria used in the overall MCA assessment indicates that generally, any slight increase in weight of a specific criterion has an impact on systems that already have a high performance on that criterion whereas in the case of systems with low performance the change is minimal or even negative. Therefore, programmes for improvement of sanitary systems might be directed to improvement options where systems already have a relatively high performance. However, those with a low performance may need comprehensive or even system reconfigurations for a significant impact to be realised.

7.6 Shifting centralised-decentralised paradigm to modernised mixtures

This section and the following Section 7.7 answer research Question 4: to what extent does the MM approach provides a useful tool for prescribing solutions and generalising on sanitary mixtures? In this section the emphasis is on the solutions, while 7.7 reflects on the usefulness of the MM approach as conceptual model.

From the MCA results in Figures 6.1-6.4 and Table 6.3, the six systems have their strong and weak sustainability elements. Based on our current insights, the way to enhance the sustainability of existing sanitary mixtures is to shift away from the centralised-decentralised paradigm, aiming for modernised sanitary mixtures (MSM) as characterised by the MM approach. Such a shift will result in merging the strengths of a centralised approach, e.g. economies of scale, efficiency, convenience; with that of a decentralised approach, e.g. accessibility, flexibility, participation, and reuse and recovery in development of intermediate systems configuration. Practically speaking, for attaining modernised sanitary mixtures in Kampala and Kisumu the following steps come to the fore:

7.6.1 Avoiding pumping stations and siphons

Avoiding pumping stations and siphons can be achieved through a shift from centralised, large-scale and mechanised systems to a mix of medium and small scale systems. The shift is imperative in reducing operation and maintenance costs through adoption of gravity based sewer systems, which can be achieved first through catchment approach in development of urban sanitary systems. This will address the poor performance of urban systems on exposure to downstream users, energy consumption, flexibility to political changes, and sensitivity to irregular maintenance (Table 6.3). Development of Eastern sewerage catchment in Kisumu, for instance, led to abandonment of Martin's Dyke and Nairobi Road pumping stations, and thereby reduced operation and maintenance costs and eliminated overflow menace. Analysis shows that once the proposed sewerage catchments are implemented fully in Kampala (NWSC, 2004), it will result in lower overall expenditure on sewerage and sewage treatment beyond the next thirty years, with power costs due to sewage pumping falling by 80% compared to current theoretical pumping costs, assuming all existing pumping stations operate as required.

However, further division of catchments in Kisumu from two to three, will result in doubling the number of pumping stations from three to six. Although costs will be reduced by 2%, it does not address the cause of poor performance to urban systems (Table 6.3). This can be different if sewage treatment in Kisumu can further be decentralised into mini collection catchments by converting pumping stations in suitable areas into STPs. Besides, the pumping stations catchments are from neighbourhood scale and above (Table 3.6). Their high sensitivity to political changes and irregular maintenance in urban systems (Table 6.3) means that pumping stations and siphons pose environmental health threats due to overflow of untreated sewage from pumping stations or blocked siphons. The proposed shift to catchment and mini-catchments will be more sustainable in terms of operation and maintenance. The local environment will benefit because the system will be operated without sewage overflows, and will help safeguard pollution of Lake Victoria.

It is apparent from Table 3.4 that some parishes within the proposed sewerage catchments in Kampala cannot economically be sewered through a catchment approach since they are too small nor can they wait for the planned period for other parishes contiguous to them to reach requisite density. Such areas, for instance, parishes like Kibowa, Kibuye II, Makindiye II, and Najjanankumbi in Nalukolongo catchment in Kampala can be suitable candidates for mini-catchment or a satellite approach as well as small-bore sewers with localised post treatment, which can be integrated into catchment sewerage and STPs when they are developed later.

7.6.2 Adoption of multiple service levels

Increasing flexibility and accessibility of sewerage systems, urban and satellite (Figure 6.2; Table 6.3), may entail a shift from exclusive application of conventional and technocratic approach in planning and design to multiple service level planning that entails a range of options: conventional sewerage, alternative sewerage and traditional onsite sanitation. In septic tank zones, settlements where the carrying capacity threshold has been exceeded, settled sewers and lower tariff rates may provide better impetus for their upgrading. The current approach of direct connection to sewer lines and uniform sewerage charges is typical of a conventional approach. Besides, analysis shows that the discounted costs over ten years resulting from by-passing the septic tank and connecting to a new sewer is much higher than if the property remains on a septic tank (NWSC, 2004). In areas with shallow rock, shallow sewers are more cost-effective than conventional sewers or onsite sanitation (MWI, 2008b). In peri-urban slums, sewer connection is possible through condominal sewers (Mara & Alabaster, 2008; UNCHS, 1986). In Kisumu, a mix of conventional sewers, condominial sewers and onsite systems (Table 3.7) are envisaged, with the former for planned and high-water consumption zones and the latter for the unplanned slum settlements

(JICA, 1998; LVSWSB, 2005a, 2008). Multiple service levels, consequently, offer differentiated sanitary options that can fit local contexts variables.

7.6.3 Modernising sewage treatment plants (STPs) to resource recovery plants (RRPs)

Shifting STPs as waste centres to resource recovery centres is aimed at addressing the poor performance of urban and satellite systems to recycling and reusing of organic matter, nutrients, energy and water (Table 6.3). This can be achieved through a shift from pure wastewater treatment processes to treatment systems that also target resource recovery. The appropriateness of STPs not only depends on the chemical water quality parameters of the treated effluent, but also on whether they meet pathogen discharge requirements, reduce operation and maintenance costs, reduce land size, curb ground water pollution and enhance resource recovery; elements that contribute to a high performance in public and environmental health criterion (Table 6.3). Also, large STPs have economy of scale advantages for resource recovery since they treat high continuous flows of organic matter and nutrients, with potentials for energy generation from biogas and nutrient recovery, especially P and N, which in principle, can be valorised as fertilizer for sale. Installation of biogas plants to convert faecal sludge into biogas and subsequent conversion of biogas into electricity is option to further explore. Moreover, rural-urban migration trends are increasing, meaning more shift and concentration of recoverable nutrient in cities. Nutrients should be ploughed back for peri-urban and rural agriculture at higher and more commercialised scales.

7.6.4 Servicing households from intermediate level

Shifting the centralised-decentralised paradigm to modernised mixtures in onsite sanitary systems can be achieved through first, establishment of intermediate faecal sludge emptying and collection infrastructures to service informal settlements to complement conventional vacuum tankers that cannot access these areas. This will address the poor performance of onsite systems to soil and water pollution and pathogen removal (Table 6.3). Vacutug and manual pit emptying technology (MAPET) are being explored as possible solutions in Kampala and Kisumu (LVSWSB, 2008; NWSC, 2004, 2008). Vacutug comprises of a tank, a small pump for extracting sludge, an easy to manoeuvre wheel vehicle, and powered by small petrol engine (Parkinson and Quarter, 2008). MAPET comprises of 200 l tank, with vacuum created by hand pump mechanism, hose and handcart (Muller and Rijnsburger, 1992). Vacutug (Mark 1) and MAPET are applied in cities of Nairobi and Dar es Salaam respectively. Application of intermediate faecal sludge emptying and collection technologies can transform illegal and insanitary manual emptying to hygienic and acceptable service providers in informal settlements. Second is to create different faecal sludge service provision levels, e.g. conventional cesspool providers in planned and accessible settlements, MAPET/Vacutug providers in informal settlements, and LAs provision of and haulage from transfer stations, coupling with MAPET/Vacutug service providers. Third, establishment of faecal sludge treatment plants (FSTPs) to enhance resource recovery and reusability of bio-solids since co-treatment with combined sewage complicates sewage treatment and may negatively impacts the quality (Table 3.3). Separate treatment of faecal sludge has some advantages as it is concentrated (Table 5.3), thus suitable for anaerobic digestion, with potentials for biogas production/energy recovery to be used locally. In addition, in concentrated streams, recovery of nutrients, N and P, is more feasible possibly attracting commercial consumers for reusing these nutrients, and thus assisting the faecal matter collection and treatment chain. Besides, trends (Figures 4.2, 5.2, 5.3) show that onsite sanitary solutions will remain dominant beyond the next two decades, with concomitant increase in faecal sludge production that needs to be collected and treated (Figure 5.6). Moreover, onsite sanitation is not transient as is often thought, but can be a permanent solution (Chapter 5). This is also reinforced by their high accessibility and flexibility (Table 6.3). Thus acceleration of onsite sanitation under the paradigm of modernised mixtures may significantly contribute to achieving the MDG of halving the number of people without improved sanitation by 2015 or WHO/UNICEF sanitation for all by 2025. Fourth, promotion of exhaustible latrines, e.g. by lining of TP and VIP latrines that are dominant (Figure 5.1), siting of shared sanitation such that they are within exhaustible distances based on average length of sucking hose pipe in the market, and regulation of emptying practices. Fifth, nutrient recovery from urine can be further explored through promotion of urine diverting toilets, where storage space is available and population is concentrated, e.g. in schools, coupled with reuse linkages, e.g. peri-urban farms or city recreation spaces.

7.6.5 Combining service provision institutions

There are multiple service providers in sanitary services provision with different strategies, payment mechanisms, management styles, and partnership arrangements (Table 7.2; Figure 7.1). In terms of institutional accessibility and resultant monopolies, urban systems are accessible to public authorities; satellite systems to state, endowed public and private institutions, and private companies; whereas onsite systems are accessible to households, private entrepreneurs, and civil society (NGO/CBOs). Access by LAs and sewerage authorities to onsite sanitary provision is limited to pilot projects. Therefore, there are no fit for all institutional models that can be adopted in a sanitary mixtures context. Applying a MM approach to institutional diversity is aimed at combining the diversified provision arrangements to fit differentiated technical and spatial scales. There are three ways to do this. First is treating households not as service consumers alone, but as providers in onsite systems, household and pay toilets. The literature portrays the service provision arrangement as a triad: public, market and voluntary sector institutions, with partnerships in-between (Picciotto, 1995; Cohen & Peterson, 1999, Blair, 2001; Glasbergen et al., 2007; Classen, 2009; Tukahirwa et al., 2010). However, as can be seen in Table 7.2 and Figure 7.1, households can be seen as the fourth category of potential service providers. Therefore, this thesis has introduced households as the fourth dimension in institutional models for service provision, coupled with multiple pluralistic institutions (Figure 7.1). A second way is combining monopolistic and pluralistic institutions to provide or deliver sanitary services in areas and scales that they have comparative advantage in. Findings in Kampala and Kisumu abound that urban systems can be provided, considering requirement for separation of asset ownership from management, through flexible public arrangements as public-public or public-private partnerships. In slum areas, public pilot sanitary projects are being implemented by NGOs/CBOs since the latter are better integrated in informal slum settlements than the public institutions. Local companies and small entrepreneurs can also compete for contracts to plan, design, build and operate such facilities. The technocratic approach by public agencies, e.g. standardised services, fixed payments, and minimal community participation, are not tenable in informal slum settlements, but are made flexible by being implemented by NGOs/CBOs. NGOs/CBOs have developed capacity and networks for participatory approaches as well as mobilisation, training, and capacity building techniques for such areas. In NGOs or CBOs developed sanitary facilities, operation and maintenance is at community level through marketised CBOs, which is a move from normal welfare to business approach to voluntary sector sanitary services provision. One configuration that needs a more flexible management arrangement is the satellite systems. Through the introduction of sanitary charges, the appointment of sewerage authorities and opening up the systems for further connections, its contribution to MDG and 'sanitation for all' targets can be enhanced. Private sector, i.e. companies and local small entrepreneurs can play a role in the operation and maintenance of decentralised satellite systems. A third way of combining provision arrangement is the regulation of onsite and faecal sludge service providers through service contracts in designated service zones. Inclusion of a users' chain, which utilises the sludge-bound nutrients for agricultural production, may stimulate the entire sanitary provision.

7.6.6 Establishing sanitary suitability areas and systems

Establishing sanitary suitability areas and systems can be achieved in two ways, use of base flow and population density thresholds (Tables 3.4 and 3.5; Figures 5.4 and 5.5). Threshold levels, though contestable, provide a general framework for mapping out present and future different sanitary areas for different systems (Figure 7.3). Utilisation of base flow and population density thresholds are supported by Kampala Sanitation Strategy and Master Plan, which stipulates that settlements with more than 200 P/ha and 10 m^3/d^*ha densities should be sewered (NWSC, 2004). The draft Sewerage Manual for Kenya (MWI, 2008b) stipulates that settlements with population densities of 120 P/ha and above be sewered, except in shallow soils where 110 P/ha is the sewer threshold. Kampala sanitation feasibility report proposes variable saturation densities for low, medium and high-income zones of 50, 250 and 450 P/ha, respectively, in line with water consumption (NWSC, 2008). Sinnatamby (1983) noted that 160 P/ha is the critical threshold for adoption of simplified sewerage in Natal Brazil instead of onsite sanitation; whereas Fang (1999) puts the density for sewerage in Indonesia at 250 P/ha. The population density at which this transition takes place varies with the physical conditions of the settlement, such as soil permeability and topography (Sinnatamby, 1983; UNCHS, 1986; MWI, 2008b). The degree of 'informality' of a settlement is also a criterion. Informal settlements such as slums have a much higher population densities, but are not sewered and are seemingly not part of the plan, which makes household onsite and conventional sewerage not tenable cost effectively. Community sanitation blocks and simplified sewerage seems to offer better impetus in such circumstances.

Establishing suitable sanitary areas and systems can also be derived from analysis and impending on city spatial structure. Based on carrying capacity (Table 6.4), urban, satellite and bio-latrines are solutions for high density areas. From Table 7.1 it can be assessed that urban sanitary systems are applied in urban core and planned peri-urban areas where connection is feasible, whereas in planned peri-urban and rural suburbs, satellite system can be applied, especially where land sizes are large. In dense peri-urban slums, bio-latrines have proved its resilience as best choice, with acceptable sensitivity to settlement type (Tables 6.3 and 6.4). Connecting the bio-digester outlet

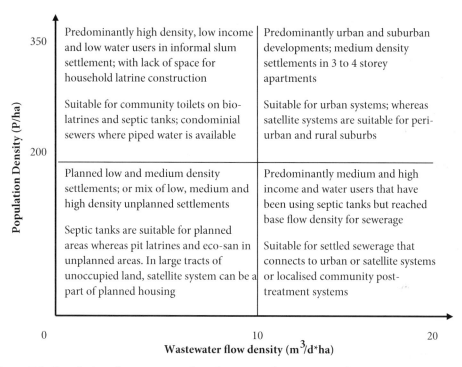

Figure 7.3. Population & wastewater flow density as determinant of sanitary systems to apply (modified from NWSC, 2004).

to nearby sewers, e.g. via a settled sewer system, would significantly enhance the sustainability of the bio-latrines as local discharge of pathogenic and nutrient rich effluents is then prevented. In low density peri-urban and rural areas, pit latrines and eco-san are suitable. Septic tanks can be part of low density urban and peri-urban housing since they have the same high convenience and low exposure to users as sewer systems (Table 6.3). Hence they are an acceptable form of sanitary solution for the middle and high-income urbanites that occupy planned low density areas. Besides, they have low requirement for institutional support, they are flexible to political changes and do not require end-user awareness (Table 6.3). Moreover, septic tanks have shown spatial concentration in Kampala and Kisumu as they are applied almost exclusively in Nakasero and Milimani, respectively. Consequently, they need to be desludged and this requires operational costs, with infiltration of effluent not always possible. If density will increase out-lets can be connected to small bore sewers and a transformation to up-flow septic tanks will further improve the onsite treatment efficiency and minimise the desludging costs. Pit latrines and eco-san can be part of low density peri-urban and rural areas.

7.6.7 Organising faecal sludge service provision into zones

Zoning technique can be applied to organise sanitary service providers and infrastructures in the city into service zones. The city can be zoned into faecal sludge collection service zones. The zones can be used to award different service contracts to different service providers depending on their comparative advantage, e.g. accessibility, flexibility, personnel capacity, and equipment they poses. This way, informal settlements can be serviced by Vacutug and MAPET and planned settlements by vacuum tankers and service providers will be awarded contracts to operate on certain zones. The more flexible requirements will force service providers to associate in order to win contracts, thus shifting the decentralised paradigm to a MM paradigm. Zoning, therefore, can be used as an instrument for curbing high operation costs inherent in cesspool tankers criss-crossing the city in search for customers and enhancing hygienic conditions through regulation of emptying, collection and disposal. Zoning and service contracts call for city councils to divest itself from service provision and focus on regulation, enforcement of compliance to hygienic standards, licensing service providers, monitoring and enforcement of contracts and development of transfer stations and modernised FSTPs.

7.7 Appropriateness of modernised mixtures approach as an assessment tool

This section attempts to answer the other part of research question 4: is the MM approach a useful assessment tool for assessing sanitary mixtures in East African cities?

Since the MM approach is an assessment tool, it can be applied anyway and is thus a generic tool. Every African city is characterized by a certain degree of patchwork of sanitary mixtures, whereas every type of sanitary system can be upgraded to meet the MM sustainability criteria.

The findings from Kampala and Kisumu show that sanitary services provision are characterised by mixed sanitary mixtures (MSM) exhibiting different configurations. MSM have different scales, which are supported by different spaces (Table 7.1) and institutional arrangements (Table 7.2) in tandem with rationalities driving them. None of the sanitary systems out-compete the other in city landscapes of East Africa, nor are they sustainable (Figure 6.1, 6.2 and 6.3). From Tables 1.1 to 1.3, the spatial, sanitary, and socio-economic structures of cities in East Africa are not significantly different. From Tables 1.1 and 1.2, the sanitary systems may not be sustainable considering the findings in Figures 6.1, 6.2, 6.3 and Table 6.3.

The MM approach has been used to assess sanitary systems configurations in Kampala and Kisumu in Chapter 3 to 6; which is espoused in Table 7.1 and Figures 7.1 and 7.2. Suffice to this is development of assessment indicators and rating scales for sanitary systems along the four MM dimensions of scale, management, flows and participation as done in Chapter 2, Section 2.5.4. The assessment tool developed can be applied in any city in East Africa as they exhibit sanitary mixtures (Tables 1.1-1.3). From present analysis, six MSM are discernible: (1) centralised urban sewerage, (2) satellite sewerage, (3) septic tanks, (4) pit latrines, (5) eco-san, and (6) bio-latrines. Bio-latrines and eco-san, which are relatively recent sanitary options in the East African landscape, are shared schemes. We assume that the assessed systems apply across East African cities. As an assessment tool, MM has demonstrated that sanitary configurations can be conventional, traditional or mixed. Besides, the assessments are not restricted to local conditions or site specific factors. They are generic and can apply to sanitary systems in any East African city or elsewhere. This thesis has demonstrated that the MM approach can also be utilised as a prescriptive tool in Chapter 6 and Section 7.6 above, as was shown in the assessment scoring the various systems on sustainability indicators of public and environmental health, accessibility and flexibility (Table

6.3). However, this is only the first part of actualising the MM approach, namely the description of the MM contexts, assessing sanitary systems, mapping sanitary configurations, and defining their boundary conditions. The second step should entail development of a mathematical model for MSM based on the boundary conditions and configurations espoused in this thesis. The third step should be to operationalize the conceptual and mathematical model through statistical survey data to generalise for developing and transition economies.

7.8 Conclusion

Sanitary provision in East African cities is rather a mixture of spatial-technical and institutional dimensions. The proposed MM approach is based on the premise that in an East African context, and implicitly in other developing countries with similar socio-economic and spatial structures, sanitary provision will be rather a mixture, comprising of different technical and spatial scales, multiple service providers and diverse institutional arrangements. Such mixtures, however, ought to be sustainable based on MM criteria of public and environmental health, accessibility and flexibility to attain sustainable urban development and meet the MDG of halving the number of people without improved sanitation by 2015 or the WHO/UNICEF Sanitation for All goals by 2025.

Sanitary mixtures are theorised as the co-existence of different phases of modernity in tandem with local context variables. Therefore, there is no one-fit-all paradigmatic way to sanitary provision if the local contexts like spatial structures, socio-economic conditions and level of environmental infrastructure development are apparently different even within the same city. However, a shift of the centralised-decentralised dichotomy to MM paradigm offers better impetus in enhancing public and environmental health, accessibility and flexibility of sanitary mixtures as it merges the strengths of centralised and decentralised approaches.

The MM approach is helpful in conceptualising, assessing, mapping and prescribing sanitary systems in cities where sanitary mixtures are the norm rather than the exception. It is also very helpful as a conceptual model for organising a research agenda which can be along the four MM dimensions: scale, management, flows and end-user participation as well as in searching for appropriate intervention pathways along one or more of the MM dimensions. It is helpful in understanding not only the scope and nature of modernisation debates, but also contextualising modernities in sanitary provision. As an assessment and decision making tool, it is helpful in finding out which elements highlighted in the assessment need to be restructured and which need improvements in order to be sustainable. However, more research is needed, towards further theoretical elaboration of the MM model and refinement of the assessment tool.

References

Abbott, J. (2010). Urban planning, physical infrastrcture and natural resources: exploring the relationship in an African urban context.Workshop Nachhaltige Infrastruktur-und Raumentwicklung. Technische Unversität Darmstadt, November 4-5, 2010.

Abu-Ghunmi, L. (2010). Characterization and treatment of grey water: options for (re)use. PhD thesis, Wageningen University, Wageningen.

Action, M.B. (1927). Sewerage Scheme-Kisumu. Kenya National Archieve ref PC/NZA/3/9/1/2, Nairobi.

Allaire, J.R. (1961). Neighbourhood Boundaries Chicago, American Societyof Planning Officials Information Report No. 141.

Anyumba, G. (1995). Kisumu town: history of the built form, planning and environment Housing and Urban Studies. PhD thesis, Delft University of Technology,Delft.

American Public Health Association (APHA) (1992). Standard methods for the examination of water and wastewater analysis. APHA, Washington, D.C.

Arnstein, S.R. (1969). A Ladder of Citizen Participation. *Journal of the American Planning Association* 35(4): 216-224.

Arts, B., P. Leroy, & van Tatenhove, J. (2006). Political modernisation and policy arrangements: a framework for understanding environmental policy change. *Public Organiz Rev* 6: 93-106

Ashley, R., & Hopkinson, P. (2002). Sewer systems and performance indicators – into the 21st century. *Urban Water* 4 123-135.

Athi Water Services Board (AWSB) (2005). Water supply and sewerage services demand forecast for Nairobi city 2005-2030. Nairobi, Athi Water Services Board (AWSB).

Bayliss, K. (2003). Utility privatisation in Sub-Saharan Africa: a case study of water. *The Journal of Modern African Studies* 41(04): 507-531.

Bijker, W.E. (1995). Of bicycles, bakilites and bulbs. Cambridge, MIT Press.

Blair, H. (2001). Institutional pluralism in public administration and politics: Application in Bolivia and Beyond. *Public Admin. and Dev.* 21: 119-129.

Budds, J., & McGranahan, G. (2003). 'Are the debates on water privatization missing the point? Experiences from Africa, Asia and Latin America.' *Environment & Urbanization* 15(2): 87-113.

Carlesen, J., Vad, J., & Otoi, S.P. (2008). Kampala City Council – A project for promoting ecological sanitation in Kampala, Uganda. Sida Evaluation 2008:44. Swedish International Development Cooperation Agency(Sida), Stockholm.

Castells, M. (1996). The rise of the network society. Oxford, Blackwell.

Castro, J.E. (2008). Neoliberal water and sanitation policies as a failed development strategy: lessons from developing countries. *Progress in Development Studies* 8(1): 63-83.

Chaggu, E.J. (2004). Sustainable environmental protection using modified pit latrines. PhD thesis, Wageningen University, Wageningen.

Chartzis, K. (1999). Designing and operating storm water drain system: emperical findings and conceptual developments. In O. Coutard (ed), *The governance of large technical systems (pp. 73-90).* London, Routledge:.

Claassen, R.J.G. (2009). Institutional pluralism and the limits of market. *Politics Philosophy Economics* 8(4): 420-447.

Cohen, J.M., & Peterson, S.B. (1999). Adminstrative Decentralization. Strategies for developing countries. Kumarian Press, Connecticut.

Cosser, P.R. (1982). Lagoon algae and the BOD test. *Effluent and Water Treatment* (September): 357-360.

Crites, R. and Tchobanoglous, G. (1998). Small and Decentralized Wastewater Management Systems. WCB – McGraw-Hill, Boston.

Crown (2009). Multi-criteria analysis: a manual. Department for Communities and Local Government. Communities and Local Government Publications, London.

Dar es Salaam Water and Sanitation Authority (DAWASA) (2008). Sanitation strategic framework for dar es salaam. Dar es Salaam Water and Sanitation Authority and Dar es Salaam City Council, Dar es Salaam.

De Graaf, P. (2006). Waterzuivering dichter bij huis – decentrale sanitatie en hergebruik. WEKI uiteverij, Amsterdam.

Dunmande, I. (2002). Indicators of sustainability: assessing the syitability of a foreing technology for a developing economy. *Technology in Society* 24(4): 461-471.

Eisenstadt, S.N. (2000). Multiple modernities. *Daedalus* 129(1): 1-29.

Ellege F. Myles, F.R., & Warner, D.B. (2002). Guidelines for the assessment of national sanitation policies. Bureau for Global Health, United States Agency for International Development Strategic Report 2, Washington DC.

El-Shafai, A.S., El-Gohary, F.A., Nasr, F.A. & van der Steen N.P. (2007). Nutrient recovery from domestic wastewater using a UASB-duckweed pond system. *Bioresource Technology* 98: 798-807.

Ertsen, M. (2005). Prescribing perfection: emergence of an engeering irrigation design approach in the Netherlands East Indies and its legacy 1830-1990. PhD thesis, Delft University of Technology, Delft.

Fang, A. (1999). On-site sanitation – an international review of World Bank experience. UNDP-World Bank, Water and Sanitation Program – South Asia.

Franceys R., Pickford, J. & Reed, R. (1992). A guide to the development of on-site sanitation. Water, Engineering and Development Centre (WEDEC), Loughborough University of Technology, UK. WHO, Geneva.

Gaye, M., & Diallo, F. (1997). Community participation in the management of the urban environment in Rufisque (Senegal). *Environment & Urbanization* 9(1): 9-30.

Gikas, P., & Tchobanoglous, G. (2009). The role of satellite and decentralised strategies in water resources management. *Journal of Environmental Management* 90: 144-152.

Glasbergeren, P., Biermann, F. & Mol, A.P.J. (eds) (2007). Partnerships, governance and sustaianble development: Reflection on theory and practice. Edward Elgar, Cheltenham.

Gómez-Ibáñez, J.A. (2008). Private infrastructure in developing countries: lessons from recent experiences. The International Bank for Reconstruction and Development/World Bank, Washington D.C.

Graham, S., & Marvin, S. (2001). Splintering urbanism: networked infrastructures, technological mobilities and the urban condition. Routledge, London.

Grau, P. (1996). Low cost wastewater treatment. *Water Science and Technolology* 33(8): 39-46.

Gray, D.E. (2004). Doing Research in the Real World. Sage, London.

Guest, J.S., Skerlos, S.J., Daigger, G.T., Corbett, J.R.E., & Love, N.G. (2010). The use of qualitative system dynamics to identify sustainability characteristics of decentralized wastewater management alternatives. *Water Science and Technolology* 61(6): 1637-1644.

Gunatilake, H., & Jose, M.J.F.C (2008). Privitization revisited: lessons from private sector participation in water supply and sanitation in developing countries. ERD Working Paper No. 115. Asian Development Bank, Manila.

Hajkowicz, S., & Haggins A. (2008). A comparison of multiple criteria analysis techniques for water resource management. *European Journal of Operational Research* 184: 255-265.

Halcombe, R.G. (1997). A Theory of the Theory of Public Goods. *Review of Austrian Economics Journal* 10(1): 1-22.

Hancard, M. (2001). Afro-modernity: temporality, politics and the African diaspora. D. Gaonkar (ed), *Alternative modernities (pp.* 272-298). Duke University Press, Durham.

Harremoës, P. (1997). Integrated water and waste management. *Water Science and Technology* 35(9): 11-20.

Harrisson, P. (2006). On the edge of reason: planning and urban futures in Africa. *Urban Studies* 43(2): 319-335.

Hasan, A. (1990). Community groups and non-governmetal organizations in the urban field in Pakistan. *Environment & Urbanization* 2(1): 74-86.

Hasan, A. (2002). A model for government-community partnership in building sewage systems for urban areas: The experiences of the Orangi Pilot Project. Research and Training Institute (OPP-RTI), Karachi. *Water Science and Technology* 45(8): 199-216.

Hasselaar, B., De graaf, P., Luising, A., & van Timmeren, A. (2006). Integratie van decentrale sanitatie in de gebouwde omgeving, Delft University of Technology, Delft.

Hegger, D. (2007). Greening sanitary systems: An end user perspective. PhD thesis, Wageningen University, Wageningen.

Hegger, D. & van Vliet, B. (2010). End-User Perspectives on the Tranaformation of Sanitary Systems. In B. van Vliet, G. Spaargaren and P. Oosterveer (eds), *Social Perspective on the Sanitation Challenge* (pp. 203-216). Springer, Dordrecht.

Ho, G. (2005). Technology for sustainability: The role of onsite, small and community scale technology. *Water Sci. Technol.* 51(10): 15-20.

Hukka, J.J., & Katko, T.S. (2003). Water Privatisation Revisited. Panacea or pancake? IRC, International Water and Sanitation Centre, The Hague.

Hunt, J., Anda, M., Mathew, K. Ho, G., & Priest, G. (2005). Emerging approaches to integrated urban water management: Cluster scale application. *Water Science and Technology* 51(10): 21-27.

International Environmental Technology Centre (IETC) (2002). Environmental sound technologies for wastewater and stormwater management: An international source book. International Water Association (IWA), Osaka.

Jaglin, S. (2002). The right to water versus cost recovery: particpation, urban water supply and the poor in sub-Saharan Africa. *Environment & Urbanization* 14: 231-245.

Jenkins, M., & Sugden, S. (2006). Rethinking sanitation: Lessons and innovation for sustainability and success in the new millenium. Human Development Report Office Occassional Paper. United Nations Development Programme, New York.

Japan International Cooperation Agency (JICA) (1998). Kisumu Water Supply and Sewerage System. Ministry of Local Government and Japan International Cooperation Agency (JICA). Volume 2, Master Plan, Nairobi.

Joubert, A., Stewart, T.J. and Eberhard, R. (2003). Evaluation of Water Supply Augmentation and Water Demand Management. *Journal of Multi-Criteria decision Analysis* 12(1): 17-25.

K'Akumu, O.A. (2006). Privatization model for water enterprise in Kenya. *Water Policy* 6: 539-557.

K'Akumu, O.A., & Appida, P.O.(2006). Privisation of urban water service provision: the Kenyan experiment. *Water Policy* 8: 313-324.

Kaggwa, J. (1994). Land Tenure and Land Use in Kampala District. Makerere University, Kampala.

Kalbermatten, J.M., Julius, D.S., & Gunnerson, C.G. (1982). Appropriate sanitation alternatives: a technical and economic appraisal. John Hopkins University Press, Baltimore.

Kariuki, M., & Schartz, J. (2005). Small-scale private service providers of water supply and electricity: a review of incidence, structure, pricing and operating characteristics. World Bank Policy Research Working Paper 3727. World Bank, Energy and Water Department and PPIAF, Washington, DC.

Kayizzi, R. (2010). 20 Years of Economic Recovery: Housing, Construction Sector Boom. The New Vision, New Vision Newspapers, Kampala.

Kenya (1974). National Developemnt Plan 1974-1978. Government Printer Nairobi.

Kenya (1979). National Developemnt Plan 1979-1983. Government Printer, Nairobi.

Kenya (1984). National Development Plan 1984-1988. Government Printer, Nairobi.

Kenya (1986). The Public Health Act. Cap 242. Government Printer, Nairobi.

Kenya (1994). National Development Plan 1994-1996. Government Printer, Nairobi.

Kenya(1995). Commission of Inquiry on Local Authorities in Kenya: A Strategy for Local Government Reform in Kenya. Ministry of Local Government, Nairobi.

Kenya (1996). The Physical Planning Act. Government of Kenya. Government Printer. Cap 286, Nairobi.

Kenya (1997). National Developement Plan 1997-2001. Government Printer, Nairobi.

Kenya (1998). The Local Government Act. Government Printer. Cap 265, Nairobi.

Kenya (1999). Planning and Building Bill. Government Printer, Nairobi.

Kenya (2002a). National Developement Plan 2002-2008. Government Printer, Nairobi.

Kenya (2002b). The Water Act. Government Printer. No. 8 of 2002, Nairobi.

Kenya (2006). The Environment Management and Co-ordination (Water Quality) Regulations, 2006. Kenya Gazette Supplement No. 68, Legal Notice No.120. Government Printer, Nairobi.

Kenya (2011). The Urban Areas and Cities Act. No 13, 2011. Government Printer, Nairobi.

Kessides, I.N. (2004). Reforming infrastructure: Privatization, regulation and competition. World Bank, Washington DC.

Khan, M.H. (2002). State failure in developing countries and strategies of institutional reform. Draft of Paper for ABCDE Conference, 24-26 June 2002, Oslo.

Kisumu Water and Sewerage Company (KIWASCO) (2008). Water and Sanitation Sector Investment Planning (SIP). Kisumu Water and Sewerage Company (KIWASCO), Kisumu.

Kombe, W.J. (2005). Land use dynamics in peri-urban areas and their implications on the urban growth and form: the case of Dar es Salaam, Tanzania. *Habitat International* 29(1): 113-135.

Kone, D. (2010). Making urban excreta and wastewater management contribute to cities' economic development: A pardigm shift. *Water Policy* 12: 602-610.

Krishna, A. (2003). Partnerships between local governments and community-based organisations: exploring the scope for synergy. *Public Administration and Development* 23(4): 361-371.

Kujawa, K., & Zeeman, G. (2006). Anaerobic treatment in decentralised and source-separation-based sanitation concepts. *Reviews in Environmental Science and Biotechnology* 5(1): 115-139.

Lee, R.M.L. (2006). Reinventing modernity: Reflexive modernization vs liquid vs multiple modernities. *European Journal of Social Theory* 9(3): 355-368.

Lee, R.M.L. (2008). In search of second modernity: reinteepreting reflexive modernisation in the context of multiple modernities. *Social Science Information* 47(1): 55-69.

Lee, T., & Floris, V. (2003). Universal access to water and sanitation: Why the private sector must participate. *Natural Resources Forum* 27: 279-290.

Loetscher, T. (1999). Appropriate sanitation in developing countries: the development of a computerised decision aid. PhD thesis, Chemical Engineering, Queensland.Brisbane.

Lake Victoria Environment Programme (LVEMP) (2001). Initiative in the management of trans-boundary waters, Lake Victoria Environment Programme (LVEMP), Kisumu.

Lake Victoria South Water Services Board (LVSWSB) (2005a). Kisumu Water Supply and Sanitation Feasibility Report. Lake Victoria Water Service Boarb (LVSWSB), Kisumu.

Lake Victoria South Water Services Board (LVSWSB) (2008). Long-Term Action Plan: Sewerage Design Report. Lake Victoria South Water Services Board (LVSWSB), Kisumu.

Lake Victoria South Water Services Board (LVSWSB) (2005b). Service provision agreement between Lake Victoria South Water Service Board and Kisumu Water and Sewerage Company (KIWASCO), Lake Victoria South Water Service Board (LVSWSB), Kisumu.

Lake Victoria South Water Services Board (LVSWSB) (2005c). Lease agreement between Municipal Council of Kisumu and Lake Victoria South Water Service Board (LVSWSB). Lake Victoria South Water Service Board (LVSWSB), Kisumu.

Mara, D. (1996). Settled sewerage in Africa. Jonh Wiley & Sons, Chichester.

Mara, D., & Alabaster, G. (2008). A nem paradigm for low-cost urban water supplies and sanitation in developing countries. *Water Policy* 10: 119-129.

Mara, D., Drangert, J.O., Anh, N.V., Tonderski, A., Gulyas, H., & Tonderski, K. (2007). Selection of sustainable sanitation arrangements. *Water Policy* 9(3): 305-318

Mara, D. (1996). Waste stabilisation ponds: effluent quality requirements and implications for process design. Water Sci. Technol 33(7): 23-31.

Mara, D. (2008). Sanitation now: What is good practice and what is poor practice? In Proceedings of Sanitation Challenge Intetrnational IWA Conference (pp. 269-273. International Water Association (IWA), Wageningen. Mason, J. (1996). Qualitative Researching. Sage, London.

Massoud, M.A., Tarhini, A., & Nasr, J. (2009). Decentralised approaches to wastewater treatment and management: applicability in developing countries. *Journal of Environmental Management* 90.

McGranahan, G., & Satterthwaite, S. (2006). Governance and getting the private sector to provide better water and sanitation services to the urban poor. *Human settlement discussion paper series*. International Institute for Environment and Development (IIED), London.

Municipal Council of Kisumu (MCK) (2008a). The Muncipal Council of Kisumu (Water Quality and Control of Effluent) By-Laws 2008. Municipal Council of Kisumu (MCK), Kisumu.

Municipal Council of Kisumu (MCK) (2008b). The Municipal Council of Kisumu (protection and conservation of the environment) by-laws. Municipal Council of Kisumu (MCK), Kisumu.

Municipal Council of Kisumu (MCK) (2008c). The Municipal Council of Kisumu (conservancy) by-laws. Municipal Council of Kisumu (MCK), Kisumu.

Municipal Council of Kisumu (MCK) (2010). Zoning Policy and Building Standards for Kisumu Municipal Council of Kisumu (MCK), Kisumu.

McFarlane, C. (2008). Sanitation in Mumbai's informal settlements: state, 'slum'and infrastructure. *Environment and Planning A* 40: 88-107.

Metcalf and Eddy (2003). Wasteater engineering treatment and reuse. McGraw Hill, 4th Edition.

Mgana, S.M. (2003). Towards sustainable and robust domestic wastewater treatment for all citizens. PhD thesis, Wageningen University, Wageningen.

Milman, A., & Short, A. (2008). Incorporating resilience into sustainability indicators: An example for the urban water sector. *Global Environmental Change* 18(4): 758-767.

Ministry of Municipal Affairs (MMA) (1999). Public-Private Partnership – A Guide for Local Government. Government of British Columbia, Ministry of Municipal Affairs (MMA), Canada.

Ministry of Health (MoH) (1987). Sanitation Field Manaual for Kenya. Department of Environmental Health. Ministry of Health (MoH), Nairobi.

Ministry of Health (MoH) (2000). National Sanitation Guidelines. Ministry of Health (MoH), Kampala.

Ministry of Health (MoH) (2002). School sanitation latrine options: Design and construction guidelines. Ministry of Health (MoH), Kampala.

Ministry of Lands (MoL) (2008). physical planning handbook. Ministry of Lands (MoL), Nairobi.

Muga, H.E., & Mihelcic, J.R. (2008). Sustainability of wastewater treatment technologies. *Journal of Environmental Management* 88(3): 437-447.

Mutikanga, H., & Mugisha, S. (2005). A phased approach to efficiency improvement in low-income countries: the case of the National Water and Sewerage Corporation in Fort Portal, Uganda. *Water Sci. Technol* 5(3-4): 281-288.

Muller, M.S. and Rijnsburger, J. (1992). MAPET: A neighbourhood based pit emptying service with locally manufactured handpump equipment in Dar es Salaam, Tanzania. WASTE, Gouda.

Murray, A., & I. Ray (2010).Commentary: Back-end users: The unregonized stakeholders in demand-driven sanitation. *Journal of Planning Education and Research* 30(1): 94-102.

Murray, A., Ray, I., & Nelson, K.N. (2009). An Innovative sustainability assessment for urban wastewater infrastrctures and its application in Chengdu, China. *Journal of Environmental Management* 90: 3553-3560.

Ministry of Water and Irrigation (MWI) (2005). Practice manual for water supply services in Kenya. Ministry of Water and Irrigation (MWI), Nairobi.

Ministry of Water and Irrigation (MWI) (2007). The National Water Services Strategy (2007-2015). Ministry of Water and Irrigation (MWI), Nairobi.

Ministry of Water and Irrigation (MWI) (2008a). Terms of Reference for Preparing a Sewerage Tariffs. Ministry of Water and Irrigation (MWI), Nairobi.Ministry of Water and Irrigation (MWI) (2008b). Draft Practice Manual for Sewerage and Sanitation Services in Kenya. Ministry of Water and Irrigation (MWI), Nairobi.

Nawangwe, B., & Nuwagaba, A. (2002). Land Tunure and Administrative Issues in Kampala City and their Effects on Urban Development. Makerere University, Kampala.

Newman, P. (2001). Sustainable urban water systems in rich and poor cities-steps towards a new approach. *Water Science and Technology* 43(4): 93-99.

Nielsen, J.H., & Clauson-Kaas, J. (1980). Appropriate sanitation for urban areas. Cowiconsult, Copenhagen.

Lens P, Zeeman G, & Lettinga G, (eds.) (2001). Decentralized sanitation and reuse – concepts, systems and implementation (pp. 116-129). IWA Publishing, London.

Nilsson, D. (2006). A heritage of unsustainability? Reviewing the origin of the large-scale water and sanitation system in Kampala, Uganda. *Environment & Urbanization* 18(2): 369-385.

Nilsson, D., & Nyanchaga, E.N. (2008). Pipes and politics: a century of change and continuity in Kenyan urban water supply. *Journal of Modern African Studies* 46(1): 133-158.

Nkurunziza, E. (2007). Informal mechanisms for accessing and securing urban land rights: the case of Kampala. *Environment & Urbanization* 19: 509-523.

National Water and Sewerage Corporation (NWSC) (2004). Sanitation Strategy and Master Plan for Kampala. National Water and Sewerage Corporation (NWSC), Kampala.

National Water and Sewerage Corporation (NWSC) (2007). Annual Report 2006-2007. National Water and Sewerage Coporation (NWSC), Kampala.

National Water and Sewerage Corporation (NWSC) (2008). Kampala Sanitation Program: Feasibility Study. National Water and Sewerage Corporationm (NWSC), Kampala.

National Water and Sewerage Corporation (NWSC) (2009). Annual Report: 2008-2009. National Water and Sewerage Corporation (NWSC), Kampala.

Odolon, J. (1998). Particpatory approach to community empowerment in sanitation promotion. In W. Simpson-Hebert, M. and S. Wood (eds). WSSCC Working Group on Promotion of Sanitation, WHO, Geneva.

Olima, W.H.A. (1994). The land use planning in provincial towns of kenya: A case of Kisumu and Eldoret. Fachbereich Raumplanung. PhD thesis, DortmundUniverity, Dortmund.

Oosterveer, P., & Spaargaren, G. (2010). Meeting social challenges in developing sustainable environmental infrastructures in East African Cities. In B. van Vliet, Spaargaren, G. and Oosterveer, P (eds), *Social perspectives on the sanitation challenge* (pp11-30. Springer, Dordrecht.

Otis, R.J., & Bakalian, A. (1996). Guideliness for the design of of simplified sewers. In D. Mara (ed), *Low-cost sewerage*. John Wiley and Sons, Chichester.

Otterpohl, R., Albold, A., & Oldenburg, M. (1999). Source control in urban sanitation and waste management: Ten systems with reuse of resources. *Water Science and Technology* 39(5): 153-160.

Otterpohl, R., Braun, U., & Oldenburg, M. (2003). Innovative technologies for decentralised water and wastewater and biowaste management in urban and peri-urban areas. *Water Science and Technology* 48(11-12): 23-32.

Palme, U., & Tillman, A. (2008). Sustainable development indicators: how are they used in Swedish water utilities? *Journal of Cleaner Production* 16(13): 1346-1357.

Palme, U., & Tillman, A. (2009). Sustainable urban water systems indicators: Researchers' recommendations versus practice in Swedish utilities. *Water Policy* 11: 250-268.

Palme, U. Lundin, M., Tillman, A., & Molander, S. (2005). Sustainable development indicators for wastewater systems – Researchers and indicator users in a co-operative case study. *Resources, Conservation and Recycling* 43(3): 293-311.

Parkinson, J., & Quader, M. (2008). The Challenge of servicing on-site sanitation in dense urban areas: Experiences from a pilot project in Dhaka. *Water Lines* 27(2): 149-163.

Paterson, C., Mara, D., & Curtis, T. (2007). Pro-poor sanitation technologies. *Geoforum* 38: 901-907.

Perry, C.A. (1939). Housing for the machine age. *Russel Sage Foundation* 49-76.

Picciotto, R. (1995). Putting institutional economics to work. From participation to governance. World Bank Discussion Papers No. 304, World Bank, Washington DC.

Pincetl, S. (2010). From the sanitary city to the sustainble city: Challenges to institutionalising biogenic (nature's services) infrastructure. *Local Environment* 15(1): 43-58.

Pombo, J.H.R. (1996). The Colombian ASAS system. D. Mara (ed) Low-cost sewerage. John Wiley and Sons, Chichester.

Randolph, J. (2004). Environmental land use planning and management. Island Press, Washington, DC.

Reed, R.A. (1995). Sustainanble sewerage: Guidelines for community schemes. Water, Engineering and Development Centre, London.

Rees, J.A. (1998). Regulation and private participation in the water and sanitation sector. TAC baground papers No. 1. Global Water Partnership, Stockholm.

Rijnsburger, J. (1996). Scale of wastewater. *Workshop proceedings on sustainable municipal wastewater treatment systems*. 12-14 November 1996, Leusden.

Sano, J.C. (2007). Urban environmental infrastructure in Kigali city, Rwanda – Challenges and opportunities for modernised decentralised sanitation systems in poor neighbourhoods. MSc thesis, Wageningen University, Wageningen.

Sasse (1998). Decentralised wastewater treatment in developing countries. Bremen Overseas Research and Development Association (BRODA), Bremen.

Scheinberg, A., & Mol, A.P.J. (2010). Multiple modernities: Transitional Bulgaria and the ecological modernisation of solid waste management. *Environment and Planning C: Government and Policy* 28(1): 18-36.

Scheinberg, A., Spies, S. Simpson, M.H., & Mol, A.P.J. (2010). Assessing urban recycling in low- and middle-income countries: Building on modernised mixtures. *Habitat International*35(2): 188-198.

Scheinberg, A. (2011). Value added: Modes of sustaianble recycling in the modernization of waste systems. PhD thesis, Wagenignen Wageningen, Wageningen.

Schwartz, K., & Sanga, A. (2010). Partnership between utilities and small-scale providers: delegated management in Kisumu, Kenya. *Physics and Chemistry of the Earth* (Article in press).

Seghezzo, L. (2004). Anaerobic treatment of domestic wastewater in subtropical regions. PhD thesis, University,Wageningen, Wageningen.

Silva, S. A., de Olivera, R., Soares, j., Mara, D.D., & Pearson, H.W. (1995). Nitrogen removal in pond systems with different configurations and geometries. *Water Science and Technology* 31(12): 321-330.

Silverman, D. (2000). Doing qualitative research: A practical handbook. Sage, London.

Sinnatamby, G.S. (1983). Low-cost sanitation systems for urban peripheral areas in northeast Brazil. PhD thesis, Leeds University, Leeds.

Smith, H. (2005). The alternative technology movement: An analysis of its framing and negotiation of technology development. *Human Ecology Review* 12(2): 106-119.

Spaargaren, G. (2003). Sustaiable consumption: A theoritical and environmental policy perspective. *Society and Natural Resources* 16: 687-701.

Spaargaren, G., Oosterveer, P., van Buuren, J., & Mol, A.P.J. (2005). Mixed modernities: Towards viable urban environmental infrastructure development in East Africa, Environmental Policy Position Paper, Wageningen University, Wageningen.

Stewart, S. (1997). Happy ever after in the marketplace: Non-government organisations and uncivil society. *Review of African Political Economy* 24(71): 11-34.

Sundaravadivel, M., Doeleman, J.A. & Vigneswaran, S. (1999). Combined surface sewerage: low-cost option for effective sanitation in semi-urban areas of India. *Environment and Policy* 1: 181-189.

Taylor, K.K., & Parkinson, J.J. (2005). Strategic planning for urban sanitation – A 21st century development priority? *Water Policy* 7(6): 569-.

Thompson, C.B. (1931). Annual report, Kisumu Londiani District. Kisumu.

Toubkiss, J. (2010). Meeting the sanitation challenge in Sub-Saharan cities: Lessons learnt from a financial perspective. In B. van Vliet, Spaargaren, G. and Oosterveer, P. (eds.), Social perspectives on the sanitation challenge (pp 163-176). Springer, Dordrecht.

Tsagarakis, T.K.P., Mara, D.D., & Angelakis, A.N. (2003). Application of cost criteria for selection of municipal wastewater treatment systems. *Water, Air, and Soil Pollution* 142: 187-210.

Tukahirwa, J. (2011). Civil society in urban sanitation and solid waste management: The role of NGOs and CBOs in metropolises of East Africa PhD thesis, Wageningen University, Wageningen.

Tukahirwa, J., Mol, A.P.J. & Oosterveer, P. (2010). Civil Society Participation in Urban Solid Sanitation and Waste Management in Uganda. *Local Environment* 15(1): 1-14.

Uganda (1995a). Water Statute. Republic of Uganda Government Printer, Entebbe.

Uganda (1995b). National Water and Sewerage Corporation Act. Uganda Printing and Publishing Corporation (UPPC), Entebbe.

Uganda (1997a). Water Act. Republic of Uganda, Government Printer, Entebbe.

Uganda (1997b). The Local Government Act. Uganda Printing and Publishing Corporation, Entebbe.

Uganda (1999). National Environment (Standard for Discharge of Effluent to River or on Land) Regulations. Republic of Uganda, Government Printer, Entebbe.

Uganda (2000). Public Health Act Kampala. LDC Publishers Cap 281, Kampala.

Uganda (2006). The Water Act (General Rates) Instruments, 2006. Cap 152. Uganda: 315-321.

UN-Habitat (2003). Water and sanitation in the world's cities. UN-Habitat, Nairobi.

UN-Habitat (2005). Situation Analysis of Informal Settlements in Kisumu. UN-Habitat, Nairobi.

UN-Habitat (2008). The State of African Cities 2008. Naiorbi UN-Habitat.

United Nations (UN) (2006). The Millennium Development Goals Report. United Nations (UN), New York.

United Nations Centre for Human Settlements (UNCHS) (1986). The design of shallow sewer systems. United Nations Centre for Human Settlements (UNCHS), Nairobi.

United Nations Development Programme (UNDP) (2006). Beyond scarcity: Power, poverty and the global water crisis. *Human Development Report 2006*. United Nations Development Programme (UNDP), New York.

United Nations Development Programme (UNDP) (2009). Overcoming barriers: Human mobility and development. *Human Development Report 2009*. United Nations Development Programme (UNDP), New York.

Van Buuren, J. (2010). Sanitation choice involving stakeholders: A participatory multi-criteria method for draiange and sanitation system slection in developing countries cities applied in Ho Chi Minh City, Vietnam. PhD thesis, Wageningen University, Wageningen.

Van der Vleuten-Balkema, A. (2003). Sustainable waste water treatment. PhD thesis, Eindhoven Technical University, Eindhoven.,.

Van Dijk, M.P. (2008). Public-private partnerships in basic service delivery: Impact on the poor, examples from the water sector in India. *International Journal of Water* 4(3/4), 216-235

Van Haandle, A.C. & Lettinga, G. (1994). Anaerobic sewage treatment: A practical guide for regions with a hot climate. John Wiley and Sons Ltd, Chichester.

Van Lier, J. B., Vashi, A., van der Lubbe, J. & Heffernan, B. (2010). Anaerobic sewage treatment using UASB reactors: Engineering and operational aspects. In H.H. Fang. (ed), *Environmental anaerobic technology: Applications and new developments*. World Scientific, Imperial College Press, London.

Van Lier, J.B. & Lettinga, G. (1999). Appropriate Technologies for for Effective Management of Industrial and Domestiuc Wastewaters: the Decentralised Apporach. *Water Science and Technology* 40(7): 171-183.

Van Vliet, B. (2002). Greening the grid: Ecological modernisation of network-bound systems. PhD thesis, Wageningen University, Wageningen.

Van Vliet, B. (2004). Shifting scales of infrastrcture provision. In D. Southerton, Chappells H. and van Vliet, B. (eds.), Sustainable consumption: The implications of changing infrastructures of provision (pp. 67-80). Edward Elgar, Cheltenham.

Van Vliet, B. (2006). The sustainable transformation of sanitation. In J. Voß, D. Bauknecht and R. Kemp (eds.), *Reflexive governance for sustainable development* (pp. 337-354). Edward Elgar Cheltenham.

Van Vliet, B., Chappells, H. & Shove, E. (2005). Infrastructures of consumption: Restructuring the utility industries. Earthscan, London.

Vanish, A.N. & Shah, N.C. (2008). Impacts of a participatory approach to assess sustainable sewage treatment technologies for urban fringe of Surat city in India. *Water Science and Technology* 57(12): 1957-1962.

Veenstra, S., & Alaerts, G.J. (1996). Technology selection for pollution control. In *Workshop on Sustainable Municipal Wastewater Treatment Systems*. ETC/Waste Consultants, Gouda.

Verschuren, P.J.M. (2002). Case study as a research strategy: Some ambiguities and opportunities. *International Journal of Social Research Methodology*. 6 (2), 121-139

Von Münch, E., & Mels, A. (2008). Evaluating various sanitation system alternatives for urban areas by multi criteria analysis – Case study of Accra Ghana. In Proceedings of Sanitation Challenge Intetrnational IWA Conference (pp. 38-47. International Water Association (IWA), Wageningen

Von Sperling, M., & Chernicharo, C.A.L. (2005). Biological Wastewater Treatment in Warm Climate Regions. IWA Publishing, London.

Water Services Regulatory Board WASREB (2007). Lincence for provision of water services. Water Services Regulatory Board (WASREB), Nairobi.

World Commission on Environment and Development (WECD) (1987). Our Common Future. Oxford University Press, Oxford.

Weitz, A., & Franceys, R. (2002). Beyond boundaries: Extending services to the Urban Poor, Asian Development Bank, Manila.

Wikipedia (2012). Ecological sanitation. http://en.wikipedia.org/wiki/Ecologica _sanitation#cite_ref-10. Accessed 2nd June 2012

World Health Organisation (WHO) (1972). Sectoral study and national programming for community and rural water supply, sewerage and water pollution control. Recommendations on national programme for community water supply development, Report No. 2. World Health Organization (WHO), Brazzaville.

World Health Organisation (WHO) (1992). The International Drinking Water and sanitation: End of Decade Review. World Health Organization (WHO), Geneva.

World Health Organisation/United Nations International Childrens Education Fund (WHO/UNICEF) (2010). Progress on sanitation and drinking-water – 2010 Update. *Joint monitoring programme for water supply and sanitation (JMP)*. World Health Organization and United Nations Interternational Children's Education Fund, UNICEF, New York and WHO, Geneva.

WSSINFO (2008). Coverage estimates-Improved sanitation-Burundi. WHO/UNICEF. On: http://wssinfo.org.

Yin, K. (1984). Case study research: Design and methods. Hills Sage, Beverly.

Zeeman, G., Kujawa, K., de Mes, T., Hernandez, L., de Graaff, M., Abu-Ghunmi, L., Mels, A., *et al.* (2008). Anaerobic treatment as a core technology for energy, nutrients and water recovery from source-separated domestic waste(water). *Water Science and Technology* 57(8): 1207-1212.

Zeeman, G., & Lettinga, G. (1999). The role of anaerobic digestion of domestic sewage closing the water and nutrient cycle at community level. *Water Science and Technology* 39(5): 187-194.

Zhang, L. (2002). Ecologizing industrialization in chinese small towns. PhD thesis, Wageningen University, Wageningen.

Zurbrugg, C., & Tilley, E. (2007). Evaluation of existing low-cost conventional as well as innovative sanitation systems and technologies. Workpackage 3 Assessment of Sanitation Systems and Technologies. EAWAG. NETSSAF, Deliverable 22 & 23. Project No. 037099.

Appendices

Appendix 1. List of resource persons from key institutions interviewed.

Institution	Officer	Main subjects
Ministry of Water and Irrigation	• Eng Simitu, Deputy Director, Operations and Maintenance	Sewerage developmental trends, policies, strategies, investment plans
	• Eng. Kasabuli, Assitant Director, Planning and Design	Sewerage planning and design, performance and challenges
Lake Vitoria South Water Services Board	• Eng. Petrolina Ogut, Chief Manager Technical Services	Sewerage development, targets challenges and operationalisation of Water Act 2002 in Kisumu
	• Eng. Agwanda, Asset Manager	Water sector transfer plan, leasing and service provision agreements
Kisumu Water and Sewerage Company (KIWASCO)	• Technical Manager Eng. Awiti (2007/8) and Eng. Jura (2008/9)	Historical development, operation and maitenance, performance, challenges, minimum service levels, future scenerio and strategies
	• Mr. Obura, Sewerage Network superindendent	Overflows, blackages and construction on top of sewers
	• Mr. Amayo, Kisat STW foreman	Operation and challenges of Kisat STW and Nyalenda WSPs
	• Mr. Anthony, Zonal Manager	Zoning and zonal operations
National Water and Sewerage Corporation	• Mr. Richard Oyoo, Chief Analyst	Performance of Bugolobi STW and satellite ponds
	• Eng. Sonko Kiwanuka, Water Production and sewerage Manager	Historical development, operation and maitenance, performance, challenges, minimum service levels, future scenerio and strategies
Kisumu Muncipal Council	• Mr. Ayani, Chief City Planner	Spatial planning and infrastrcture development
	• Ms. Belinda, Public Health Officer	Sanitation regulations, practices and challenges
Kampala City Council	• Kampala Urban Sanitation Project (KUSP) ofifcer	Rationale, success and challenges of KUSP
	• Kampala Ecological Sanitation (KESP) officer	Rationale, success and challenges of KESP
	• Public Health Officers, Kawempe, Central and Nakawa Divisions	Sanitation regulations, practices and challenges

Satellite providers and operators	• National Housing and Construction Company, Arkright and NSSF personnel	Satellite areas characteristics; planning, design & management; challenges and future plans
Cesspool Emptiers	• Mr. Matovu Jafari, General Secreatry	History, operations, charges, association, challenges and prospestcs
Non-Governmetal Organisations NGOs	• SANA International Director (Kisumu); UWASNET Secretary General (Kampala)	Role of voluntary sector in sanitation provision, challenges, success and future senario
Umande Trust	• Technical Manager (Mr. Francis) • Director 9 (Mr. Omotto)	Bio-latrine technology Location and end-user particpation

Appendix 2. List of persons participated in sustainability performance assessment.

Name of expert	Institutional affiliation	Country
Mr. Richard Oyoo	National water and Sewerage Corporation	Uganda
Mr. Joseph Kirabira	Kampala Capital City Authority	Uganda
Ms. Selima Rajab	Odongo and Odongo Company Limited	Kenya
Mr. Fredrick Salukele	Environmental Engineering Department, Ardhi University of Dar es Salaam	Tanzania
Prof. S. Mbulingwe	Environmental Engineering Department, Ardhi University of Dar es Salaam	Tanzania
Dr. Gabor Szanto	PROVIDE Project, Wageningen University	Netherlands

Summary

The urbanisation of poverty and informality in East African cities poses a threat to environmental health, perpetuates social exclusion and inequalities, and creates service gaps (UN-Habitat, 2008). This makes conventional sanitation provision untenable citywide, giving rise to the emergence of sanitation mixtures. Sanitation mixtures have different scales, institutional arrangements, user groups, and rationalities for their establishment, location, and management. For assessing the performance of both the mixtures as a whole and the different sanitation approaches constituting these mixtures, novel approaches for analyses are required. This thesis, therefore, departs from the centralised-decentralised approaches to a modernised mixtures (MM) approach in seeking a more inclusive assessment of sanitary configurations taking into account public and environmental health, accessibility and flexibility of sanitation systems as sustainability criteria. To achieve this, the four objectives formulated for this thesis are to:

1. Make an inventory of sanitary systems in Kampala and Kisumu.
2. Assess and map sanitary systems along MM dimensions in Kampala and Kisumu.
3. Assess sustainability of sanitary systems on defined MM criteria in Kampala and Kisumu.
4. Enhance insights on the applicability of MM criteria as conceptual model, assessment and prescriptive tool for sanitary mixtures in East African cities.

Case study cities were chosen from a typology of primary and secondary cities that have urban sewer systems since colonial times. The two cities were deemed to offer rich cases that would give a general outlook of other East Africa cities, thus can offer possibilities for generalization. The thesis utilised a multi-method and multi-level approach in data collection and analysis. A multi-criteria analysis is used in sustainability performance assessment of sanitation systems based on defined MM criteria.

Firstly, modernisation debates and resultant modernities in sanitation provision were reviewed in Chapter 2. The review shows that Western modernisation and resultant modernities and their structures of service provision have not resonated well in developing countries. Consequently, alternative theories that dispute a universal approach to modernity emerged to offer alternatives to modernisation. Alternative options are characterised by multiple rationalities, diversity and multiplicity. Modernities in terms of sanitation provision are further operationalized as competition between the proponents of centralised versus decentralised solutions. A third way of looking at sanitation modernisation that is more inclusive is advanced through the introduction of the MM approach.

In Chapter 3, the presence of urban sewer systems in Kampala and Kisumu cities is assessed. The results show that urban systems are of medium scale and serve about 10% of the city population. They are publicly owned and managed by public enterprises under new public management. Besides, they are conventionally designed, constructed and operated without the involvement of end-users. Treatment plants are either overloaded, underutilised or treatment stages are mismatched. Consequently, about 30-70% of the treatment stages are not operational. Effluent discharge standards and bio-solids reuse requirements are not met, and the adopted treatment technologies are inappropriate for the investigated conditions. Sewer networks are supported by pumping stations and siphons that are only partially operational due to high operational costs and mechanical failures. Public sewerage is further plagued by urban informality and multiplicity of city spatial structures. Planned city core, and to a limited extent peri-urban areas, are served by public sewers, while sewer trunk lines pass through informal slum settlements without connections.

In Chapter 4, satellite systems are analysed and configurations mapped. Satellite systems are intermediate semi-collective decentralised sewerage and treatment systems developed parallel to urban and onsite systems. They are provided by multiple actors, serve planned middle and high income residential, industrial complexes, endowed public and private universities, and government facilities. In terms of scale, they are community, neighbourhood and small-urban sanitation solutions. Besides, satellite systems are private sewerage systems that utilise gravity sewers and localised mechanised or non-mechanised treatment. The flows are based on land use or facility specific and are treated close to the point of generation. They are based on conventional designs and construction protocols without end-user involvement.

Onsite systems in Kampala and Kisumu cities are examined in Chapter 5. Planning forecast indicates that onsite systems will dominate sewer (urban and satellite) systems beyond the next two decades. They are small-scale, highly decentralised and use simple technologies. Pit latrines dominate septic tanks in number, with eco-san on pilot scales and bio-latrine being a new sanitation option. Faecal sludge collection, treatment and safe disposal is dismal. The private sector dominates over local authorities in provision of faecal sludge services, but public sewerage agencies receive and co-treat faecal sludge with sewage although sewage works are not designed to receive faecal sludge. They are regulated by the Ministry of Health, enforced by the city councils and are provided by multiple actors solely or in partnership. Onsite sanitation can be a transient or permanent solution depending on mass flows and spatial requirements. However, for better sanitation provision, a permanent solution, with room for amendments to anticipate changes in space and mass flow is imperative.

In Chapter 6, sustainability performance of sanitation systems are assessed following the defined three MM criteria. The performance shows that there is no sanitation system that is completely outcompeted in performance, neither are there systems with a very good performance. Sanitation system choices, consequently, are made among imperfect options, which call for balancing the various elements of sanitation provision to suit different policy and local contexts. Varying the assigned relative weight of the various criteria used in the overall MCA assessment indicates that generally, any slight increase in weight has an impact on systems that already have a high performance whereas in the case of systems with low performance the change is dismal or even negative. Therefore, programmes for improvement of sanitation systems might be directed to improvement options where systems already have a relatively high performance. However, those with a low performance may need comprehensive or even system reconfigurations for significant impacts to be realised.

In conclusion, sanitation mixtures are theorised as the co-existence of different phases of modernity in tandem with local context variables. Thus, there is no one-fit-all paradigmatic way to sanitation provision if the local contexts are apparently different even within the same city. However, a shift of the centralised-decentralised dichotomy to modernised mixtures paradigm offers better impetus as it can utilise the advantages of both centralised and decentralised approaches without jeopardising existing provision pathways. The MM approach is helpful in assessing, mapping and describing sanitation systems in cities where sanitation mixtures are the norm rather than the exception.

One way to modernise sanitation mixtures is by shifting the centralised-decentralised paradigm in order to modernise the mixed sanitation landscape. This is premised on the notion that such a shift will result in merging the strengths of centralised approach, e.g. economies of scale, efficiency, and convenience, with strengths of decentralised approach, e.g. accessibility, flexibility, participation, and reuse and recovery in development of intermediate systems configuration. This can be achieved through, among others, avoiding use of pumping stations, adoption of multiple service levels, involvement of private sector, servicing households at intermediate scale, and establishing sanitation suitability and management zones.

The MM approach is also very helpful as a conceptual model for organising a research agenda which can be set along the four assessment dimensions of scale, management, flows and participation, as well as in searching for appropriate intervention measures along one or more of these dimensions. As an assessment and decision making tool, it is helpful in finding out which elements highlighted in the sustainability assessment need to be restructured and which need improvement in order to enhance their sustainability. However, translation of the proposed conceptual MM model into a mathematical model is a challenge yet to be explored. Considering its intrinsic dynamic character in dependence to varying spaces, flows, and scales along city development, a mathematical MM model would provide a regulatory design tool for city planners for adopting amendments to existing sanitation solutions. Obviously, up to date monitoring and inventory records are a pre-requisite for applying such a model, requiring institutional upgrading. Although the current results described in this thesis provide the basis for a more structured assessment and generalisation of sanitation mixtures, more research and contextualisation is needed in other regions, for further elaboration of MM model, and for the refinement of the assessment tool.

Curriculum Vitae

Sammy C. Letema was born on 23 September 1971 in Elgeyo-Marakwet County, Kenya. He went to Chebai Primary School from 1980 to 1987. In 1988 to 1991, he pursued his Kenya Certificate of Secondary Education in Tambach High School. He then joined Kenyatta University for a Bachelor of Environmental Studies (Planning and Management) in 1993-1997 and Master of Environmental Planning and Management in 2000-2002. From 2003-2004 he worked briefly as a consultant in formulation of municipal strategic plans and integrated solid waste management plans. From 1995 he has lectured environment and spatial planning in the Department of Environmental Planning and Management, Kenyatta University. In 2007, he joined Wageningen University for his PhD Studies sponsored by INREF under PROVIDE project of Wageningen University. He followed his PhD programme under SENSE graduate school. Results of his PhD has been presented in international conferences and published in peer reviewed journals and book chapters. His PhD work won an innovative award price in 2011 from 5 S-NL Society of The Netherlands for the purpose of drafting a mathematical model on sanitation selection based on the modernised mixtures approach.

Printed in the United States
by Baker & Taylor Publisher Services